Phyllotaxis and Symmetry in Angiosperms

Phyllotaxis and Symmetry in Angiosperms

Jean-Paul Walch
Solange Blaise

iSTE

WILEY

First published 2024 in Great Britain and the United States by ISTE Ltd and John Wiley & Sons, Inc.

ISTE Ltd
27-37 St George's Road
London SW19 4EU
UK

www.iste.co.uk

John Wiley & Sons, Inc.
111 River Street
Hoboken, NJ 07030
USA

www.wiley.com

Library of Congress Control Number: 2024935368

British Library Cataloguing-in-Publication Data
A CIP record for this book is available from the British Library
ISBN 978-1-78630-989-1

Contents

Foreword

The understanding of phyllotaxis in plants has been of major interest to botanists for a very long time and remains a hot and still unresolved topic of discussion. Additionally, phyllotaxis and symmetry are intrinsically linked, as transitions between spirals and whorls are essentially responsible for the floral diversification among angiosperms. Symmetry in flowers, especially the causes of shifts between different types of symmetry, follows specific patterns that are ruled by developmental and genetic factors.

Solange Blaise and Jean-Paul Walch have recently arrived at the forefront of botanical research, but their contributions have already delivered great impact. A comprehensive book on phyllotaxis models was published in 2023 and has sown the seeds for this new book on symmetry in flowers and its links with phyllotaxis.

The authors' approach goes beyond that of traditional botany, adding their expertise in mathematics and physics to deal with fundamental questions on floral morphology. The book clearly demonstrates the intimate relationship between phyllotaxis and symmetry in flowers. The authors explore this connection from different angles, including experimental models, genetics and physics. They also discuss different types of symmetry and how these relate to major groups of plants. The law of symmetry is represented by continuous growth processes, demonstrating the intricate relationships between organs in time and space, leading to a wonderful pattern in the sequential differentiation of inflorescences and flowers.

The authors explore patterns of symmetry in major groups of angiosperms, citing several studies of specific species related to the causes of changes in symmetry patterns. The book is well illustrated with clear examples of symmetry and their causes taken from other works and is beautifully redrawn. The book emphasizes the fundamental link between growth and form, highlighting the importance of undertaking developmental studies. As such, they revisit the contributions of the great floral morphologists of the 20th and early 21th century but with a novel, contemporary approach. Most of the observations and conclusions are based on published studies of the early development of flowers, as young stages provide a much better picture of the processes affecting symmetry. Understanding the kind of symmetry necessitates a combination of different approaches, including evidence of spatial constraints and genetic shifts. Through a series of clear examples, the authors provide a detailed overview of the difference between early ("constitutional") and the late ("positional") zygomorphy. They clearly demonstrate the intricate relationship of genetic stimuli and spatial constraints in shaping the floral structure and demonstrate how important it is to approach flowers from different angles.

This book will contribute to our understanding of the intricate mechanisms shaping flowers and should find a space on the bookshelf or computer of anyone interested in the structure of flowers and their evolution.

Louis Ronse De Craene
Royal Botanic Garden
Edinburgh

Preface

> There cannot possibly be a physical or dynamical, though there may well be a mathematical Law of Growth, which is common to, and which defines, the spiral form in the Nautilus, in the Globigerina, in the ram's horn, and in the disc of the sunflower.
>
> The essential facts of [Phyllotaxis] are not difficult to understand, but the theories built upon them are so varied, so conflicting and sometimes so obscure, that we must not attempt to submit them to detailed analysis and criticism.
>
> In the morphology of living things, the use of mathematical methods and symbols has made slow progress, and there are various reasons for this failure to employ a method whose advantages are so obvious in the investigation of other physical forms.
>
> "On Growth and Form", D'Arcy Thompson (1917)

D'Arcy Thomson suggested that there could be a mathematical law common to the spirals of plants and animals. We have discovered this law: it is a law of symmetry, in the same way that radial and bilateral symmetry are. This law allows us to bridge phyllotaxis and floral symmetry, and explains why purely physical phenomena, in particular magnetic, also generate spirals.

Furthermore, it was necessary to identify the physical and biochemical process which ensure the implementation of this law in plants. Using a unified system of equations, we have modeled the molecular mechanisms and pressure forces that act on inflorescence and floral meristems. Thus, mathematical methods allow us to gain insight into what gives flowers their shapes: our model is an instrument for evaluating the forces on meristem from their effect on the geometry of this meristem. With this powerful tool, we have revisited many important notions of plant biology.

Spiral phyllotaxis optimizes the packing of primordia (e.g. sepals and petals on the floral meristem), but there is no clear boundary between organ types: thus, organ number is labile, as seen in early-diverging angiosperms. Packing is less optimal in whorled phyllotaxis; however, organs within a whorl have the same identity and therefore each species has a well-defined number of organs of each type. Magnoliales and Laurales use a "strategy" to align organ identity with spiral phyllotaxis based on "pseudo-whorls", that is, sets of organs of the same type that look like whorls but are less regular because they are derived from Fibonacci spiral initiation (with divergence angles between successive primordia equal to the golden angle (137°), e.g. quincuncial aestivation) as opposed to rational angles dividing the circle into equal parts (e.g. 72° = 360°/5 in a pentamerous flower). Each pseudo-whorl corresponds to the periphery of the apex of the meristem where Fibonacci spirals are initiated, which explains why the number of organs in a pseudo-whorl is a term of the Fibonacci sequence (3, 5, 8, 13, etc.). This is why monocots are trimerous and dicots are often pentamerous.

Inflorescences themselves display obvious symmetrical patterns (e.g. distichous, opposite-decussate or spiral). These patterns are transmitted to the perianth via the bracteoles in cymose inflorescences and a few racemose inflorescences, or by contact pressure in the absence of these transitional organs. In a flash of genius, Eichler (1875, 1878), in his monumental work on floral diagrams, identified the rules of the influence of bracteoles on the arrangement of the calyx, but despite his efforts, he was unable to give these rules a "mathematical" explanation. This is what our modeling has been able to do.

We show that although tetramerous perianths can be radially symmetrical from initiation, in general, pentamerous perianths are initiated as irregular pentagons. The transition to regular pentagons (radial symmetry: for example, valvate aestivation) requires an equalization of the pentagon angles before anthesis.

If spirals have remained the preserve of phyllotaxis, the study of the genetic causes of floral symmetry has taken off since the characterization of floral symmetry genes in *Antirrhinum majus* (Luo et al. 1996). More than 100 studies using molecular methods (RT-PCR and RNA in situ hybridization) have shown that closely related members of the *TCP* gene family play a crucial role in the control of floral symmetry in diverse angiosperm lineages.

By transposing the conclusion of Curie's (1894) thesis on electromagnetic phenomena, stating that "when certain causes produce certain effects, the elements of symmetry of the causes must be found in the effects produced", to the symmetry of flowers, we were able to state a general principle whereby the symmetry of the perianth (the effect) is derived from the symmetry of certain *TCP* gene expression (the cause).

The model developed in our previous book, *Phyllotaxis Models*, by Jean-Paul Walch and Solange Blaise made it possible to identify the general mechanisms responsible for the location of organ primordia within the floral meristem. In this book, we highlight the interplay between the symmetry of the expression of certain *TCP* genes and previously described phyllotactic mechanisms, be they of molecular inhibition or mechanical (contact pressure) origin.

Endress (1999) defined two categories of zygomorphy, constitutional (apparent as early as the initiation of perianth primordia) and positional (appearing late in floral development), and asked: what is the genetic difference between these two categories? We show that these two categories of zygomorphy correspond to two different temporalities of floral asymmetrical *CYC*-like gene expression: early or late. Late zygomorphy is caused in particular by the development of a spur (e.g. *Delphinium* (Ranunculaceae)) or by elongation of the flower (ligules of Asteraceae).

May 2024

Acknowledgments

We would like to thank Florian Jabbour, Associate Professor at the Muséum National d'Histoire Naturelle, Paris, for his encouragement, Hélène Citerne for her review of the manuscript, and Line and Michèle Walch for their superb drawings.

Introduction

In our previous book, *Phyllotaxis Models*, by Jean-Paul Walch and Solange Blaise, we developed a mathematical model of phyllotaxis that made it possible to reproduce in silico the arrangement of the floral organs in dozens of species; these patterns matched the SEMs (scanning electron micrographs) of developing flowers, as well as the floral diagrams of Eichler (1875, 1878). In this book, we studied a number of species from angiosperm orders with mainly bilaterally symmetrical flowers from a phyllotactic and genetic point of view.

Angiosperm orders are not presented following the APG IV (2016) phylogenetic system but rather a pedagogical order. However, in each chapter, families within an order follow the APG IV classification system. Species were selected according to the availability of data for phyllotaxis modeling: SEMs, floral diagrams of Eichler and drawings of plant organogenesis (Payer 1857). The expression patterns of floral symmetry genes were compiled from reverse transcriptase polymerase chain reaction (RT-PCR) and RNA in situ hybridization data. Chapters 3–7 are therefore made up of case studies.

In Chapter 1, we give readers the keys to understanding this "new phyllotaxis", which is one of the two pillars of our approach to floral symmetry. The other pillar is the genetics of floral symmetry. In particular, we focus on *CYC*-like genes from the *TCP* family that are implicated in the control of floral symmetry in angiosperms.

Chapter 2 is the only chapter to use mathematical concepts. We show that phyllotactic spirals are a form of symmetry. If Fibonacci spirals are the most common, with the number of spirals corresponding to numbers in the Fibonacci sequence (3, 5, 8, etc.), other configurations show spiral numbers corresponding to the Lucas sequence (3, 4, 7, 11, etc.). We show that the latter are the consequence of dislocations within the lattice of primordia.

In Chapter 3, we model floral organogenesis in Ranunculales (early-diverging eudicots) and analyze the genetics of floral symmetry in species where data are available. The flower of bleeding heart (*Lamprocapnos spectabilis*, Papaveraceae) presents an unusual symmetry (disymmetry), which corresponds to a disymmetric expression of a *CYC*-like gene. The bilateral symmetry of *Capnoides sempervirens* (Papaveraceae) is linked to the growth of a lateral spur (the single axis of symmetry is transverse). Again, the tepal arrangement is consistent with the symmetry of *CYC*-like gene expression. In the actinomorphic corolla of *Aquilegia* (Ranunculaceae), *CYC*-like gene expression is itself radially symmetrical. These genes are not expressed in the actinomorphic corolla of *Nigella* (Ranunculaceae). *Delphinium* is one of the rare genera in Ranunculaceae with zygomorphic flowers (associated with the development of a large spur): its *CYC*-like genes present a late asymmetric expression along the dorsoventral axis. As in many Ranunculaceae, the perianth of *Ranunculus repens* is initiated in a spiral. We show how the arrangement of the perianth shifts toward actinomorphy, due to contact pressure between organs during growth.

Antirrhinum majus, from the order Lamiales, is the model plant for the genetics of floral symmetry. We study the perianth of six Lamiales with either actinomorphic or zygomorphic flowers by simulating their floral phyllotaxis using our model and showing the influence of *CYC*-like gene expression patterns on the phyllotaxic mechanisms (Chapter 4). In accordance with Curie's principle, the symmetry of the expression of these genes determines the symmetry of the perianth.

In Chapter 5, we distinguish constitutional zygomorphy (early zygomorphy, e.g. Lamiales) from positional zygomorphy (late zygomorphy). For example, in *Tropaeolum longifolium* (Brassicales), zygomorphy is not the consequence of the arrangement of floral organs during their initiation, but of their differential growth, in particular, with the formation of a spur. In *Impatiens* (Ericales), the weight of the spur causes the flower to rotate at anthesis (resupination). In Asteraceae, the causes of the actinomorphy of disc florets and the zygomorphy of ray flowers are also consistent with Curie's principle.

Unlike in most core eudicots, the first sepal in Caesalpiniodeae and Faboideae (Fabales) is generally located on the abaxial side of the flower, which requires that the adaxial inhibition source be much stronger than that of the bract, in the vicinity of which this first sepal arises. Modeling a zygomorphic perianth requires that couples of primordia appear simultaneously and symmetrically with respect to the dorsoventral axis ("twinning"), for example, initiation of twin petals forming the two keels or the two wings, as in *Cercis*. In *Mimosa* (Mimosoid clade), the symmetry of the inflorescence (a raceme), as well as the synchronous development of the flowers within it, underlies the radial symmetry of these flowers. In *Pisum*

(Faboideae), with typical papilionoid flowers, the symmetry of *CYC*-like gene expression in the flower is the cause of zygomorphy (Chapter 6).

Most early-diverging monocots are actinomorphic, while more derived monocot families, such as Orchidaceae, Iridaceae, Poaceae, the order Zingiberales and Commelinaceae, are mainly zygomorphic. Curie's principle applies just as well to the perianths of monocots as to those of eudicots. Therefore, in Commelinaceae, *TCP* genes are not expressed in the actinomorphic flowers of *Tradescantia pallida* but are expressed asymmetrically with respect to the transverse axis in the zygomorphic flowers of *Commelina communis*. These two species are a perfect illustration of scorpioid cymes, the former having a bracteole, while the latter does not (Chapter 7).

The Conclusion presents the most significant outcomes of our theory.

Our model has been published in the *Journal of Theoretical Biology*, its application in basal angiosperms in *Botany Letters* and the demonstration of the influence of bracteoles on sepal arrangement (initially intuited by Eichler in 1875) in *Flora*. The influence of *CYC*-like genes on phyllotaxic mechanisms in Lamiales has been published in *Botany Letters*.

1

Concepts of Phyllotaxis and the Genetics of Floral Symmetry

1.1. Types of floral symmetry

In the generally accepted classification system, four types of floral symmetry are defined (Endress 1999) (Figure 1.1):

– *Actinomorphy* (radial symmetry or polysymmetry): the flower has n axes of symmetry, where n > 2 is the number of organs in each whorl (e.g. six axes of symmetry; Figure 1.1(A)).

– *Disymmetry*: there are two perpendicular axes of symmetry (Figure 1.1(B)).

– *Zygomorphy*: (bilateral symmetry or monosymmetry): the flower is symmetric with respect to the dorsoventral axis and asymmetric with respect to the transverse axis (Figure 1.1(C)).

– *Asymmetry:* flowers have no axis of symmetry (Figure 1.1(D)).

Bilateral floral symmetry is predominant in certain species-rich families such as Lamiaceae and Fabaceae (legumes) in eudicots and Orchidaceae in monocots. Phylogenetic studies have demonstrated that zygomorphy has evolved many times from ancestors with radially symmetrical flowers and that reversals from bilateral to radial symmetry are frequent. A minimum of 130 shifts from radial to bilateral floral

For a color version of all the figures in this chapter, see www.iste.co.uk/walch/phyllotaxis2.zip.

symmetry and 69 reversals to actinomorphy have been inferred within angiosperms. For example, in Lamiales, bilateral symmetry has evolved early in the diversification of the clade and has been followed by at least 10 reversals to radial symmetry (Reyes et al. 2016).

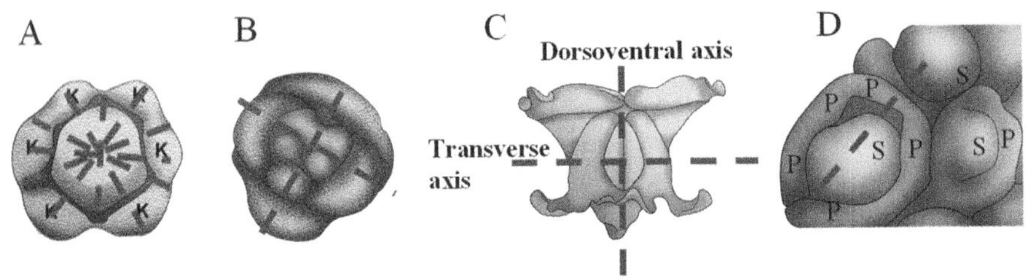

Figure 1.1. *Different types of floral meristem symmetry before anthesis. (A) Actinomorphy: Antirrhinum majus (Plantaginaceae) cyc:dich mutant at an early developmental stage (from Luo et al. 1996) with six axes of symmetry. (B) Disymmetry: Lamprocapnos spectabilis (Papaveraceae) with two perpendicular axes of symmetry (from Damerval et al. 2013). (C) Zygomorphy: wild-type Antirrhinum majus (redrawn from Vincent and Coen 2004). Dashed lines represent the dorsoventral and transverse axes, which intersect at the center of the flower. (D) Flower meristems of Centranthus ruber redrawn from Roels and Smets (1996). The asymmetric flowers have a single stamen that is not on the dorsoventral axis. K = sepals from the German Kelch, not to confuse with S = stamen; P = petal; S = stamen*

In flowers that have spur, the third dimension has to be taken into account, and symmetry is considered with respect to planes instead of axes (Figure 1.2(A)). In the case of *Impatiens* (Balsaminaceae; Figure 1.2(A)), the weight of the spur tilts the flower at 180° at anthesis (resupination).

> *Jean-Baptiste Payer* was appointed professor of geology and mineralogy at the University of Rennes in 1840. In 1844, he obtained the chair of botany at the École Normale Supérieure in Paris. He received a doctorate in medicine from the Paris faculty in 1852 and was appointed that same year as professor of plant organography at the faculty of sciences. He published "On Comparative Plant Organogeny of the Flower" between 1854 and 1859.

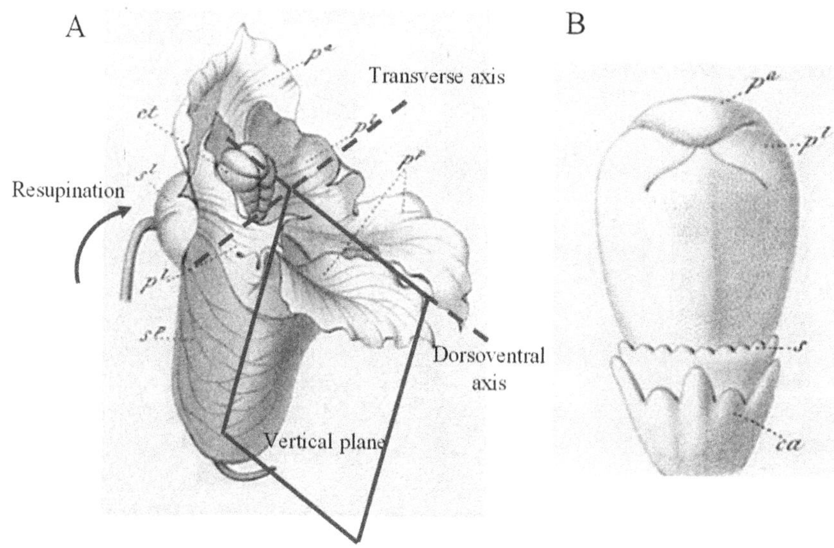

Figure 1.2. *(A) Impatiens glandulifera (Balsaminaceae) and (B) Dipsacus laciniatus (Caprifoliaceae) from Payer's "Traité d'organogénie végétale comparée de la fleur" (1857). General legend from Payer: fl = flower; ca = calyculus; s = sepal; p = petal; et = stamens (étamines); ep = spur (éperon); superscript: a = anterior; l = lateral; p = posterior. Dashed blue lines = axes of symmetry; straight lines = plane of symmetry*

Box 1.1. *Drawings of developing and mature flowers by Jean-Baptiste Payer*

1.2. The bracteole theory

In 1875, the genius of botany A. W. Eichler studied at length the relationship between flowers and bracteoles (*Vorblätter. Anschluss und Einsatz der Blüthe* in Volume I of *Blüthendiagramme* (1875)) and described the different possible positions of the sepals. When only one bracteole is present – in particular, in monocots – the position of the first sepal/tepal (K1) is opposite the bracteole. When two bracteoles are present (α, the older bracteole, and β, the younger bracteole), Eichler considered that the sepals were located on the same spiral as the bracteoles (Figure 1.4(A)).

Eichler tried to explain sepal arrangement "mechanically" using Hofmeister's rule (1868), that is, a new primordium is initiated in the largest space available between pre-existing primordia. This rule predicts an alternation of bracteoles and sepals. However, the many exceptions to this rule forced Eichler to abandon the path opened by Hofmeister.

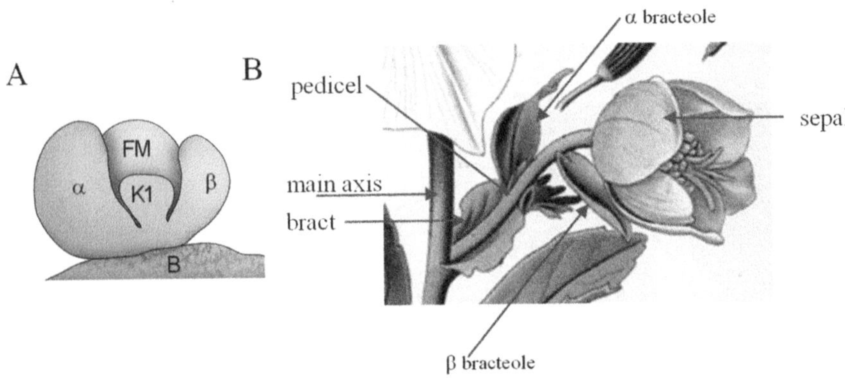

Figure 1.3. *(A) Initiation of the bract (= B), floral meristem (FM), bracteoles (α and β; the older bracteole α being larger) and the first sepal (K1) in Aquilegia vulgaris (redrawn from the SEM of Tucker and Hodges 2005). (B) Illustration of a flower of Helleborus niger (Thomé 1885). Bracteole α is below β on the pedicel*

Our model is based on the equations of Douady and Couder (1996a) that simulate the effect of an inhibition potential (Box 1.2) generated by pre-existing primordia on the location of the subsequent primordium. If we hypothesize that bracteoles determine the arrangement of sepals, this dynamical model of phyllotaxis, based on an inhibition field, makes it possible to account for the numerous arrangements (trimery, tetramery and pentamery) found in monocots and eudicots. This hypothesis suggests that determining sepal arrangement is the first biological function of bracteoles, as intuited by Eichler in 1875. Our model of phyllotaxis explains Eichler's different cases in a "purely mechanical" way, thus responding to the challenge launched by Eichler 150 years ago (Walch and Blaise 2022b).

Consider a meteoroid located 400 km away from Earth: the force of universal attraction pulls the meteoroid towards the center of our planet. In the absence of this meteoroid, something remains: it is the attraction field created by the Earth. The gravitational potential is a measure of this attraction field, and it takes a second object, like a meteoroid, for the potential force to materialize into an actual force.

Like universal attraction, the inhibition generated by a primordium decreases as the distance from that primordium increases (Figure 1.4(C)).

Box 1.2. *What is a potential?*

1.3. Phyllotaxis

1.3.1. *Phyllotaxis models*

Our phyllotaxis model and its parameterization have been described in detail in Walch (2022) and Walch and Blaise (2023). Flower organs (sepals, petals, stamens and carpels) originate from meristems as small primordia and are initiated in a crown at the periphery of the central zone (CZ), or "front", of the meristem at a certain distance from its center (Figure 1.4(A); Godin et al. 2020).

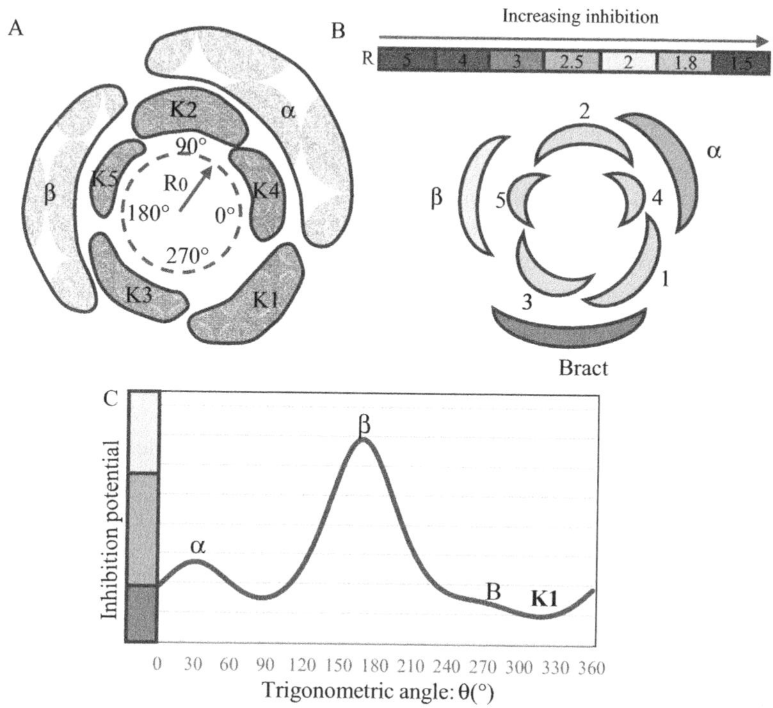

Figure 1.4. *(A) Arrangement and initiation sequence of the bracteoles and sepals in Ranunculus repens (from Meicenheimer 1979). K1 to K5 = sepals numbered from the oldest to the youngest. Dashed circle = periphery of the central zone of the meristem or "front" where organs are initiated. R_0 = radius of the front circle. Main trigonometric angles are shown. (B) Model of the calyx of Ranunculus repens. B = bract; α, β = bracteoles. Colors represent the "effective distance" (R; see Box 2.1) of the inhibition sources from the center of the flower, and the intensity of the inhibition increasing, the closer the organ producing the inhibition is to the flower center. (C) The inhibition profile along the periphery of the CZ (or "front": dashed circle in Figure 1.4(A)) at the initiation of K1. Local maxima are produced by existing primordia (here, the bract and the two bracteoles). K1 is located at the potential minimum (θ = 316°)*

Organ	Angle (°)	Distance from the center
Bract	270	3
α	30	2.5
β	168	2

Table 1.1. *Modeling parameters for quincuncial arrangement of Figure 1.4. Parameters are polar coordinates of organs*

For example, to model the perianth of *Ranunculus repens* (Ranunculaceae), whose initiation is described by Meicenheimer (1979) (Figure 1.4(A)), we considered three sources of inhibition: the bract and the two bracteoles, α and β, initiated in this chronological order; therefore, β is closer to the CZ than α (and α is closer to the CZ than the bract). The two bracteoles are located in the continuity of the vegetative spiral, and therefore have an angular distance of 138° (Walch 2022). According to the model, the inhibition potential at the front is higher the closer primordia are to the front (corresponding to younger primordia, here bracteole β). The subsequent primordium (primordium K1) will be positioned at the inhibition potential minimum at the front (Figure 1.4(C)). This first sepal (K1) will be practically opposite the most inhibitory bracteole (β) and the arrangement of the calyx will "mechanically" be quincuncial, with sepal K2 in the adaxial position (Figure 1.4(B)).

Ranunculus repens perianth phyllotaxis is "pseudo-whorled", because it looks like a whorl but is less regular as it is derived from spiral initiation with divergence angles between successive primordia more or less equal to the golden angle (137°), as opposed to rational angles dividing the circle into equal parts (72° = 360°/5 in a whorled pentamerous flower). It is therefore necessary to distinguish the **cycle** – "Cyklus" (in German) – from the whorl – "Quirl".

Generally speaking, to model the perianth of the flowers studied, we assigned values to eight input parameters: the distance from the center of the flower (R) and the angular coordinates (θ) of four "peripheral organs": an adaxial organ (the "inflorescence axis"), the bract (B) and the two bracteoles (α and β). The shorter the distance from the center, the higher the inhibition at the front.

The front of the meristem can be represented on a plane by a circle of fixed radius R_0 (Figure 1.4(A)), chosen as the unit of length. Primordia are represented as points between this circle and the outer edge of the meristem (Figure 1.7(B)). A new primordium i is initiated on this circle at a point of polar coordinates (R_0, θ), where θ is the angle that minimizes the inhibition potential generated by pre-existing primordia (Figure 1.4(C)).

In our model, primordia are produced in the following manner. The angular positions (θ) and radial distances (R) from the center (i.e. the polar coordinates) of the first sources of inhibition are given as input of the model (e.g. Tables 1.1–1.3), which makes it possible to calculate the value of the inhibition potential they produce at any point of the meristem. The first sepal primordium appears at the periphery of the central zone (or "front") at the point where the inhibition potential is at its minimum (e.g. 316° in Figure 1.4(C)). The inhibition potential at any point of the meristem is then calculated to take into account pre-existing primordia, including the newly initiated sepal. The second sepal primordium is then positioned at the periphery of the CZ, at the point where the inhibition potential generated by these pre-existing primordia is at its minimum, etc., until the positions of all organs are calculated.

In the absence of observational measurements of the distance of primordia in the species examined, we needed to find, by trial and error, the parameters that would allow us to reproduce in silico the primordia arrangement known from floral diagrams and/or SEM photographs. All studied species have quite different arrangements of their perianth. We considered that modeling was successful when the following criterion was satisfied: all primordia in the model must be located between the same neighboring primordia as on the floral diagram or SEM. This method is used in statistical physics to compute the macroscopic behavior of a system from a microscopic model (e.g. intermolecular interactions computed with the Lennard–Jones potential model).

We consider that our mathematical model is a tool for evaluating the "forces" in (i.e. biochemical factors) and on (i.e. pressure) the meristem from their effect on the geometry of this meristem (arrangement of organs).

The range of the inhibition around bracts and bracteoles may be reduced as a result of the physical discontinuity between these external organs and the floral meristem. Therefore, the distances of these primordia from the center of the floral meristem should be understood as "effective" distances and not actual distances as measured within the floral meristem (Box 2.1).

Primordia appear at regular time intervals on the front (in black in Figure 1.5(A)). The time interval between the initiation of two consecutive primordia is called the *plastochron* (T). Richards (1951) defined the *plastochron ratio* (r_i) as the distance from the apex of primodium i (R_i) over the distance from the apex of its chronological successor (R_{i+1}) (Figure 1.5(A)):

$$r_i = \frac{R_i}{R_{i+1}} > 1 \qquad [1.1]$$

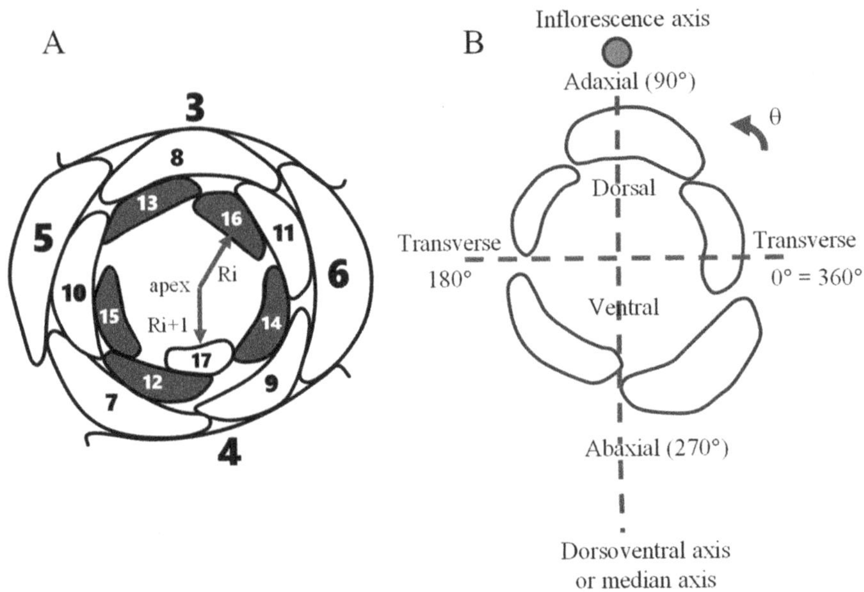

Figure 1.5. *(A) An 11-day old vegetative meristem of Linum usitatissimum (Linaceae) from Williams (1975). New leaf primordia arise on the front (numbers 12–16 in black and the youngest primordium (17) in white). Primordia are numbered from the oldest to the youngest. Ri and Ri+1 are used to calculate the plastochron ratio (see text below). (B) Generic floral diagram (see Ronse de Craene 2022), with names of flower part locations: transverse and dorsoventral axis, adaxial/abaxial side of the dorsoventral axis and dorsal/ventral parts of the flower. The main angles on the trigonometric circle (θ) are given*

For instance, if the distance from the apex of primordium 16 is equal to 77 μm and that of primordium 17 is 70 μm (as measured on a drawing of *Linum usitatissimum* by Williams 1975), then the plastochron ratio between the initiation of primordium 16 and that of primordium 17 is 77/70 = 1.1.

Since organs are in contact, high plastochron ratios are equivalent to organs being large (Figure 1.6(A)); whereas, small plastochron ratios (close to 1) are equivalent to organs being small (Figure 1.6(B)).

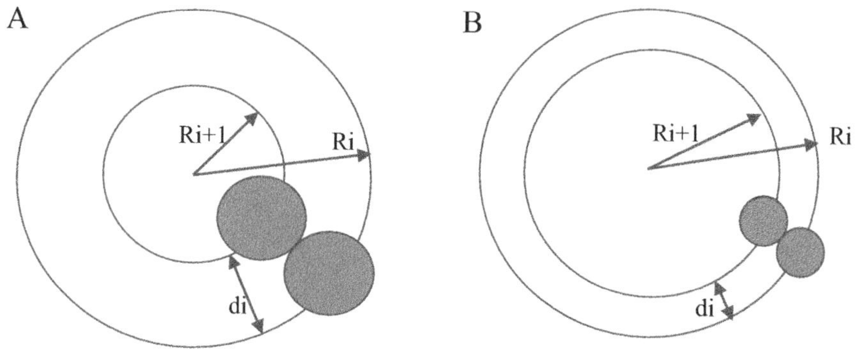

Figure 1.6. *Relationship between the plastochron ratio (R_i/R_{i+1}) and organ size (d_i): (A) a high plastochron ratio; (B) a small plastochron ratio*

1.3.2. *Fibonacci spirals*

Illicium anisatum (Schisandraceae) is commonly known as "star anise" and belongs to the ANA grade of basal angiosperms (Amborellales, Nymphaeales and Austrobaileyales), which diverged early in the history of angiosperms from the lineage that leads to all other extant flowering plants. At early stages of floral development, all organs of *I. anisatum* are spirally arranged, the divergence angles being approximately 138° (Endress 2001; Figure 1.7(A)).

Organ number	Angle (°)	Plastochron ratio	Organ number	Plastochron ratio
1	120	1.34	13	1.08
2	300	1.29	14	1.08
3		1.25	15	1.07
4		1.22	16	1.07
5		1.19	17	1.07
6		1.17	18	1.06
7		1.15	19	1.06
8		1.13	20	1.06
9		1.12	21	1.06
10		1.11	22	1.06
11		1.10	23	1.06
12		1.09	24	1.05

Table 1.2. *Modeling parameters for Illicium*

We used the Douady and Couder (1996a) equations to simulate the phyllotaxis of floral organs in *Illicium* and the "canonical" values for the parameterization of the model (see Chapter 2).

Spiral phyllotaxis are produced by a continuous and decreasing series of plastochron ratios (Table 1.2). Organs number 1 and 2 give initialization data. Next angles are calculated by the model.

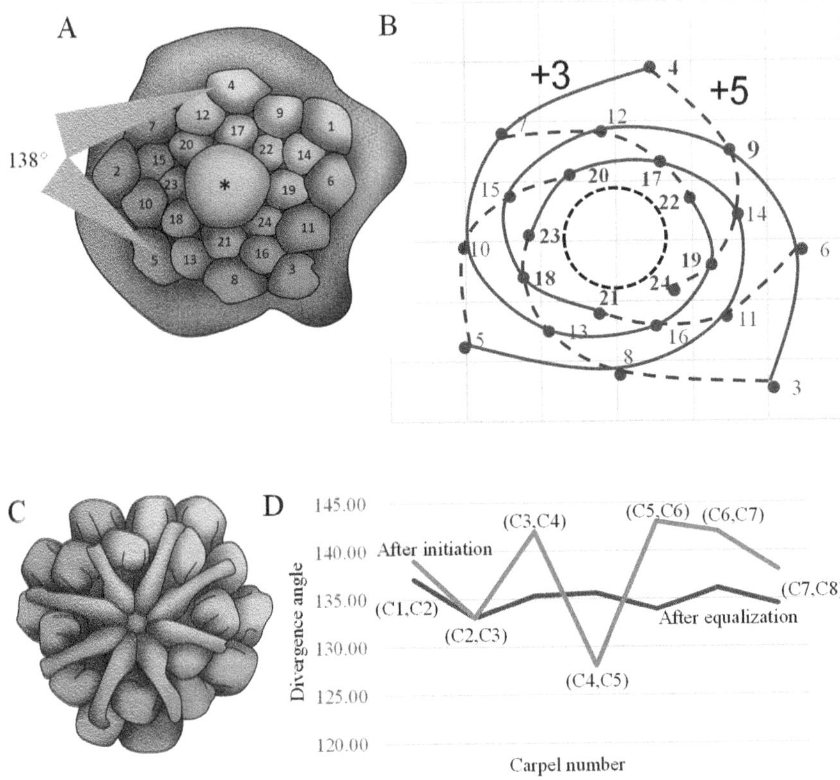

Figure 1.7. *(A) Floral organ initiation in Illicium anisatum (redrawn from an SEM micrograph from Endress 2001). Numbers indicate the order of floral organ initiation, from the oldest to the youngest. The "divergence angle" between two chronologically consecutive primordia is approximately 138°. (B) Model of the floral meristem of I. anisatum. There are three spirals (in red) turning in one direction (+3) and five spirals (dashed blue) turning in the opposite direction (+5): the "phyllotaxis mode" is (3, 5). (C) I. anisatum at anthesis (from Endress 2001) showing the star-shaped carpel arrangement. (D) Predicted divergence angles between chronologically consecutive carpels after initiation and after equalization (see text). Cn = carpel number; (Cn, Cn+1) = angle between Cn and Cn + 1*

To determine the number of "parastichies" (spirals) in a spiral structure, we can either count them or subtract the number of two consecutive primordia on the same parastichy (e.g. 9 – 4 = 5 in Figure 1.7(B)). The number of parastichies in a Fibonacci spiral structure belongs to the sequence: 2, 3, 5, 8, ... where each number is the sum of the two previous numbers. This sequence is the Fibonacci sequence, where the ratio of two consecutive terms, such as 3/2, 5/3, 8/5..., tends to the golden ratio (~1.618) as the terms increase.

If there are m spirals turning in one direction and n spirals turning in the opposite direction, the "phyllotaxis mode" is **(m, n)** (e.g. (3, 5) in Figure 1.7(B)) and the number of primordia at the front is equal to (m + n) (i.e. 8 for primordia 17–24 in Figure 1.7(B)). For example, for mode (2, 3), the number of primordia at the front is 5 (pentamery: see Figure C.11).

The angle between two chronologically consecutive primordia is called the divergence angle (Figure 1.7(A)). In the same way that ratios of consecutive Fibonacci numbers (n/m) converge towards the golden ratio, divergence angles in meristems with numbers of parastichies that are terms of the Fibonacci sequence are close to the golden angle, which is equal to:

$$\phi = (3 - \sqrt{5})/2 \times 360° \approx 0.382 \times 360° \approx 137.51° \qquad [1.2]$$

Weisse (1894) had already observed that some species with spiral phyllotaxis during development display a whorled arrangement at anthesis. We modeled this transformation in the gynoecium of *I. anisatum* (Figure 1.7) by simulating the pressure exerted on organ i by $(n - 1)$ organs (here $n = 8$ for carpels), which are approximately located on the same circle (the meristem front comprising n organs that will form the future whorl) (Walch and Blaise 2022a). Within the same whorl, the contact pressure between the different primordia tends to equalize the angles between pairs of contiguous primordia. If two of them are closer than would be expected by equal spacing (360°/n), greater pressure will tend to push them apart. If two primordia are further apart than equal spacing, less pressure will tend to bring them closer together. We call this phenomenon the "equalization process" of primordia arrangement during primordia growth.

The contact pressure model explains the transition from a spiral to a whorled arrangement of the carpels in *I. anisatum*. At the end of the recursive process, the angular distances between the primordia on the front are more or less equal (Figure 1.7(C)). The divergence angles between chronologically consecutive carpels, which are unequal after spiral initiation, all approach 135°, which means that two geometrically consecutive carpels in this whorl are separated by 45°, corresponding to a perfectly whorled arrangement (Figure 1.7(D)).

1.3.3. *Whorled phyllotaxis*

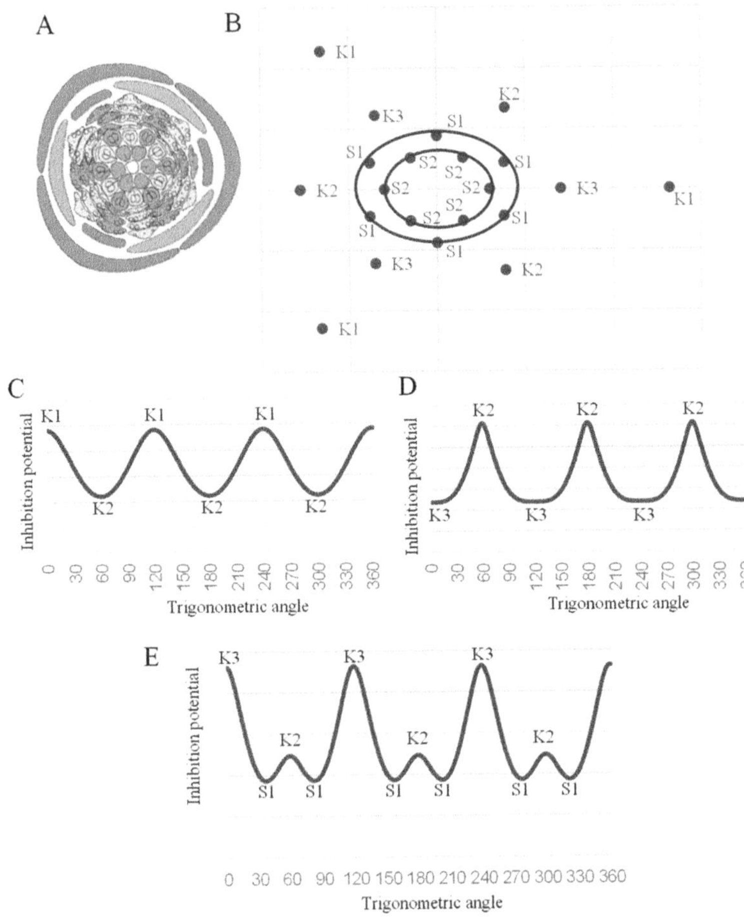

Figure 1.8. *(A) Drawing of a microtome section of floral meristem of Anaxagorea crassipetala to show the organ phyllotaxis of an entire flower just before anthesis (from Endress and Armstrong 2011). Blue and green: tepals (each color corresponds to a separate whorl); alternate stamen and carpel orthostichies are in yellow and red. (B) Model of floral organ initiation in A. crassipetala, up to the second stamen whorl (S2). Organs (K = tepals, S = stamens) are numbered according to their corresponding whorl. (C) The inhibition potential profile on the meristem front at the initiation of the second tepal whorl: the angular locations of existing maxima are located at K1; tepals K2 are located at the minima of the inhibition potential. (D) Profile on the meristem front of the inhibition potential at the initiation of the third tepal whorl: the angular locations of existing maxima are located at K2; tepals K3 are located at the minima of the inhibition potential. (E) Profile on the meristem front of the inhibition potential at the initiation of fourth whorl producing a dédoublement (see Box 1.3) of stamens in A. crassipetala*

The flower of *Anaxagorea crassipetala* (Annonaceae, Magnoliales) is composed of nine tepals, 42 stamens (of which 6–12 are sterile, i.e. staminodes) and 18 carpels. The three tepal whorls are trimerous. The first androecium whorl is composed of six stamens in "double position", alternating with the three inner tepals. The androecium consists of seven whorls of six stamens and the gynoecium of three whorls of six carpels. The androecium and gynoecium form 12 regular "orthostichies" (radial rows in red or yellow on Figure 1.8(A)) (Endress and Armstrong 2011).

Organ number	Angle (°)	Plastochron ratio	Organ number	Plastochron ratio
K1	0	1	S1	1
K1	120	1	S1	1
K1	240	1.7	S1	1.4
K2		1	S2	1
K2		1	S2	1
K2		1.1	S2	1
K3		1	S2	1
K3		1	S2	1
K3		1.2	S2	1.2
S1		1	S3	1
S1		1	...	...
S1		1		

Table 1.3. *Plastochron ratios used to model floral organ initiation in Anaxagorea crassipetala from Walch and Blaise (2023). Organs K1 give initialization data. Next angles are calculated by the model*

The process of initiating a whorl is very different from the process generating a spiral: the initial conditions (first whorl: K1) have to be perfectly symmetrical and all primordia of this whorl must be initiated simultaneously (0, 120 and 240°). Therefore, when the first primordium of the second whorl (K2) is initiated, the inhibition potential exhibits three minima with equal inhibition potentials due to the perfect symmetry of the first whorl. Therefore, three primordia will initiate simultaneously at these minima (Figure 1.8(C)). Plastochron ratios are equal to 1 within a whorl, and jump ($r > 1$) between two whorls (Table 1.3).

In *A. crassipetala*, *dédoublement* occurs in the fourth whorl (S1). According to our model, this doubling is due to the small plastochron ratio between whorls K2 and K3 ($r = 1.1$; Table 1.3), so that the inhibitory field generated by K2 is still strongly felt at the initiation of S1. S1 stamens are therefore doubled because of the combined influence of K2 and K3, which generates six minima (Figure 1.8(E)).

For Staedler and Endress (2009), the "double position" of organs in certain whorls is a striking characteristic: "two or more organs are in a position where only one is expected". It most often appears during the transition between the perianth and the androecium, but also sometimes within the perianth or the androecium, or even at the onset of gynoecium initiation. This phenomenon might be explained by the fact that doubled organs (stamens vs. petals) are smaller than those of the preceding whorl. It is a sign of great flexibility: indeed, *dédoublement* can occur at different stages of floral development or in one whorl but not in the next.

Box 1.3. *Dédoublement (Staedler and Endress 2009)*

1.3.4. *Molecular and contact pressure origin*

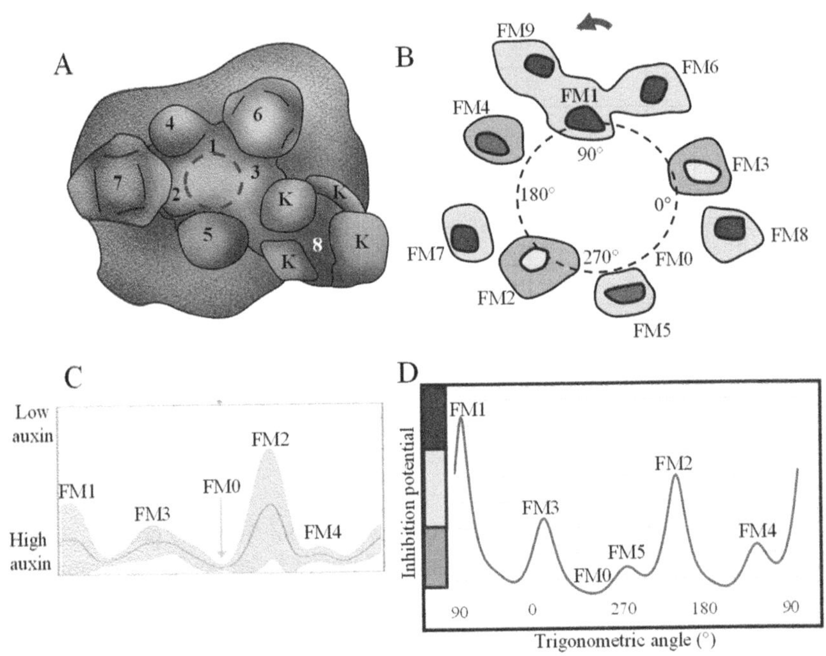

Figure 1.9. *(A) The inflorescence raceme of Arabidopsis thaliana redrawn from an SEM micrograph from Smyth et al. (1990). (B) Location of A. thaliana floral meristems according to the model. Floral meristems are initiated on the front of the inflorescence meristem (dashed line). Numbers indicate the order of floral primordia initiation, from the youngest (FM1) to the oldest (FM9). (C) Profile of the auxin reporter concentration at the front as measured by Godin et al. (2020). (D) Model of the inhibition profile at the front. FM numbers are as in (B); the next primordium to be initiated is FM0, which will be located at the minimum of inhibition profile. Color represents the inhibition potential at front generated by a primordium (Figures 1.9(B) and (D))*

In the case of the raceme meristem of *Arabidopsis thaliana*, we showed (Walch 2022) that the inhibition potential corresponded to the inverse of the auxin concentration (which is an activator), whose accumulation induces flower initiation at the periphery of the central zone, as measured by Galván-Ampudia et al. (2020) (Figure 1.9).

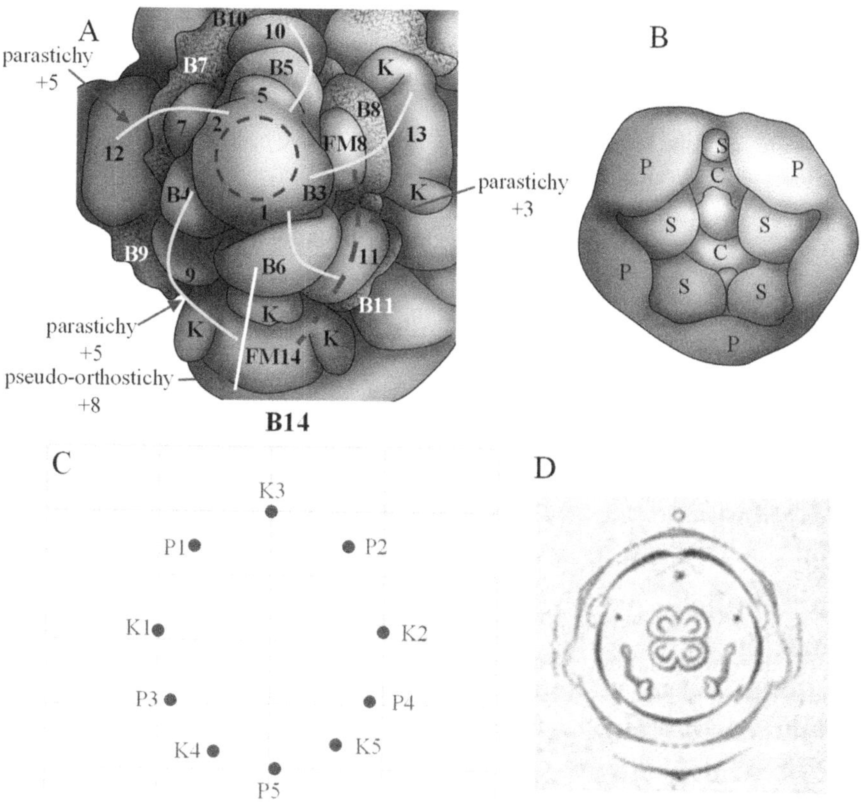

Figure 1.10. *(A) The inflorescence apex of Antirrhinum majus (redrawn from an SEM from Carpenter et al. 1995). Numbers correspond to floral meristems (FM), 1 being the youngest; Bn corresponds to the bract number; bracts B7-B10 are not shown so that floral meristems are visible. Yellow lines: the five counterclockwise parastichies (+5); red dashed line: one of the three clockwise parastichies (+3); white lines: pseudo-orthostichy (+8). (B) The floral meristem of A. majus at stage 6 redrawn from Vincent and Coen (2004). (C) Model of the floral meristem of A. majus. (D) Floral diagram of Salvia officinalis (Eichler 1875), including cryptic bracteoles. K = sepal, P = petal, S = stamen, C = carpel*

Auxin is concentrated in FM1, causing strong inhibition (in red in Figure 1.9(B)) that strongly pushes back the place of initiation of the next primordium (FM0). The latter is also pushed back, to a lesser extent, by other FMs, notably FM2 and FM3 (in yellow). The angular distance of FM0 from FM1 is thus 137°.

Bull-Hereñu et al. (2022) noted that, in racemose inflorescences, rigid bracts compress the young developing flowers. When these bracts exert extensive pressure, floral primordia become squeezed and even flattened (FM14 in Figure 1.10(A)). We hypothesized that in the absence of bracteoles, such as in *Antirrhinum majus*, the forces shaping the perianth are not due to biochemical activity but are to contact pressure from the bracts subtending the flowers above each developing flower in the inflorescence (Walch and Blaise 2022b).

Organ	**Angle (°)**	**Distance from the center**
Adaxial source (B6)	90	2
Bract (B14)	270	1.8
B11	50	2
B9	120	3

Table 1.4. *Parameters used to model the flower of Antirrhinum majus. B11 and B9 are the bracts located on the same two parastichies as the developing floral meristem (FM14); B6 is on the same pseudo-orthostichy than FM14; B14 is F14 bact (see Figure 1.10A)*

In the absence of bracteoles, we hypothesized that four sources of inhibition influence the initiation of sepals in *A. majus* (Table 1.4): the bract subtending the flower (B14 subtending FM14; θ_{B14} = 270° = absolute angle of bract B), the two bracts located above the flower on the two parastichies passing through this flower (θ_{B11} = 50°; θ_{B9} = 120°) and the large adaxial bract located on the *pseudo-orthostichy* passing through the flower (θ_{B6} = 90°) ("pseudo", because unlike in whorled phyllotaxis, it is not exactly radial). The angle of these three upper bracts was measured on the SEM photograph of Carpenter et al. (1995) (Figure 1.10(A)). The sepal initiation sequence generated by the model (Figure 1.10(C)) matches that of FM14 and the perianth model matches that of the SEMs of Carpenter et al. (1995) (Figure 1.10(B)).

In some of his floral diagrams, Eichler (1875) added *Unterdrückt* (cryptic) bracteoles, which do not form visible structures. He introduced cryptic bracteoles mainly from theoretical assumptions (discussed in Choob 2022). These are represented by dashed patterns (e.g. Figure 1.10(D)). According to our interpretation, the contact pressure from the bracts subtending the flowers above each developing flower in the inflorescence corresponds to the cryptic bracteoles of Figure 1.10(D).

1.4. Genetic control of floral symmetry

Pea (*Pisum sativum*; Faboideae) was the original model organism used in Mendel's discovery of the laws of inheritance (1866), making it the organism on which modern plant genetics is founded. Pea flowers are zygomorphic; the three petal types are arranged along the dorsoventral axis (Figure 1.11(B)). A large collection of mutants is known. For example, two loci associated with floral symmetry, *KEELED WINGS (K)* and *LOBED STANDARD 1 (LST1)*, were identified in 1919 (by Caroline Pellew, who worked on the "rogue" phenomenon in peas with William Bateson) and 1985 (Winfield 1987), respectively. Mutation at the *LST1* locus results in an abnormal shape of the dorsal petal; whereas, the *k* mutant has a homeotic transformation phenotype, that is, the lateral and dorsal petals resemble the ventral petals in size, shape and color (Wang et al. 2008) (homeotic genes are genes that regulate the development of anatomical structures. They often encode transcription factor proteins, and these proteins affect development by regulating downstream gene networks involved in body patterning).

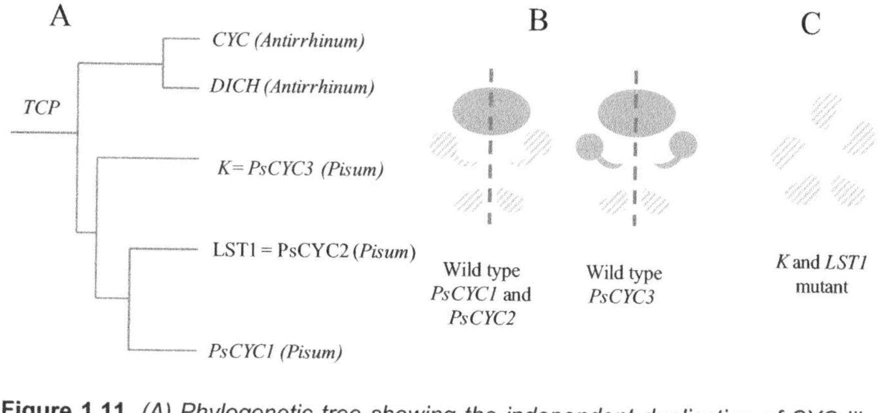

Figure 1.11. *(A) Phylogenetic tree showing the independent duplication of CYC-like TCP genes in Antirrhinum (CYC and DICH) and Pisum (PsCYC1, PsCYC2, PsCYC3) from Wang et al. (2008). (B) Expression in the corolla of CYC-like genes in Pisum sativum. Blue = expression of CYC-like genes, striped = no expression. The shape of the flower is stylized to represent the symmetry of the corolla at anthesis, as in Hileman (2014). (C) k and lst1 double mutants in Pisum lack expression in the corolla (from Wang et al. 2008)*

Wang et al. (2008) demonstrated that these loci control lateral (*K*) and dorsal floral organ identity (*LST1*) and originate from the duplication of a *TCP* gene during the diversification of Faboideae (Figure 1.11(A)). Differences between wild-type and *k* and *lst1* mutants become visible at around the same developmental stage (stage 6), when vascular tissues in petals are developed and the internally asymmetric shape of both lateral and ventral petals becomes obvious (Wang et al. 2008).

When dorsal (and lateral) regulator function is abolished, the default state appears to be a ventralized form, where all petals acquire ventral identity (Figure 1.11(C)).

In many core eudicots, the genetic mechanisms underlying floral bilateral symmetry and the differential growth of organs along the dorsoventral axis involve *TCP* (acronym for *TEOSINTE BRANCHED 1, CYCLOIDEA,* and *PROLIFERATING CELL FACTOR*) and *MYB* (Myeloblastosis) genes. In *Antirrhinum majus*, *CYCLOIDEA (CYC)* and *DICHOTOMA (DICH)*, which belong to the *TCP* gene family, are expressed only in the adaxial region of the flower (Figure 1.12(B)). They activate a *MYB* gene, *RADIALIS (RAD)*. Together, they negatively regulate cell division at early stages of development, resulting in changes of symmetry and merosity. Furthermore, *RAD* negatively regulates *DIVARICATA (DIV)*, another *MYB* gene, on the adaxial side, thus restricting its function to the abaxial region of the flower (Luo et al. 1996; Almeida et al. 1997; Corley et al. 2005; reviewed in Martinez-Salazar et al. (2021)).

Two major *CYC*-like gene duplications occurred prior to the diversification of core eudicots, resulting in three distinct *CYC*-like gene clades: *CYC1*, *CYC2* and *CYC3* (Howarth and Donoghue 2006). Within each clade, further gene duplication events occurred independently in many core eudicot lineages. For example, there are at least five *CYC2* genes in sunflower (Chapman et al. 2008). In addition, *CYC*-like gene duplications have also occurred independently in monocot and basal eudicot lineages. This is due to independent *CYC*-like gene duplication events at a small clade level (reviewed in Spencer and Kim 2018).

Functional analyses in recent years have indicated that *CYC1*, and to a lesser extent *CYC3* genes, play a central and well-conserved regulatory role in shoot branching and the development of axillary meristems; whereas, the *CYC2* genes are key players in the control of floral symmetry, inflorescence architecture and reproductive organ development (reviewed in Nicolas and Cubas 2016).

Two genes are homologous if they shared ancestry in the evolutionary history of life. They can have shared ancestry either because of a speciation event (*orthologs*) or a duplication event (*paralogs*).

Box 1.4. *CYC-like genes*

In almost all major angiosperm lineages, the key genes responsible for establishing the dorsal–ventral identity of flowers have been shown to be *homologs* of *CYCLOIDEA* (*CYC*). The genetic factors underlying bilateral floral symmetry in *Antirrhinum majus* were identified through the analysis of mutants in which zygomorphy is reduced or lost (Luo et al. 1996). Mutants with a peloric phenotype have radially symmetrical flowers in which all organs have a ventral identity, while the semipeloric phenotype is intermediate between the peloric and wild-type phenotypes. Differences between wild-type, semipeloric and peloric meristems appear when sepal primordia are initiated. In peloric and semipeloric mutants, six sepal primordia emerge (Figure 1.12(C)).

> Pelorism has been of interest since a five-spurred variety of the common *Linaria vulgaris* (toad flax; Plantaginaceae) was first described by Carl Linnaeus in 1744. Linnaeus termed this type of mutant as a "Peloria", the Greek word for "monster" or "prodigy".
>
> Mature wild-type flowers of *Linaria vulgaris* have five petals that are united for part of their length to form a corolla tube ending in five separate lobes. Zygomorphy is clearly evident in the shape and color of the petals. Flowers of naturally occurring peloric mutants of *Linaria vulgaris* are radially symmetrical. All five petals resemble the ventral petal of the wild type, each having a small lobe with an orange lip and a spur at their base. Compared to the peloric mutants of *Antirrhinum*, merosity does not change. The peloric mutation in *Linaria* is recessive, as crosses with the wild-type yield essentially wild-type F1 progeny (Cubas et al. 1999).

Box 1.5. *Peloric Linaria*

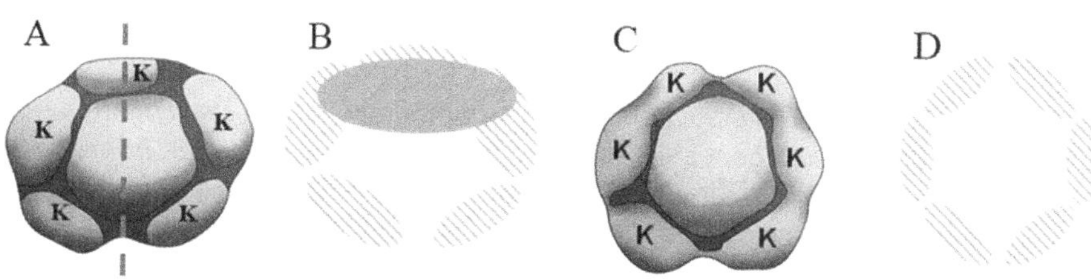

Figure 1.12. *(A) Late stage 4 (initiation of sepal (K) primordia) meristem of wild-type Antirrhinum majus redrawn from an SEM from Luo et al. (1996). (B) Schematic representation of CYC expression (in blue) in a stage 4 floral meristem of wild-type A. majus (Luo et al. 1996). (C) Late stage 4 floral meristem of hexamerous peloric A. majus redrawn from an SEM from Luo et al. (1996). (D) Lack of CYC expression in the late stage 4 peloric floral meristem. Blue = expression of CYC-like genes; striped = no expression*

In wild-type *Antirrhinum*, *CYC* expression is detected as early as stage 1 of floral development (prior to organogenesis). When sepal primordia become visible, *CYC* expression is detectable in the dorsal sepal and in the dorsal part of the central zone of the meristem (Figure 1.12(B)) (Luo et al. 1996).

In *A. majus*, mutants that lack both *CYC* and *DICH* expression have radially symmetrical peloric flowers (Figure 1.12(D)); whereas, single *CYC* mutants have semipeloric flowers and single *DICH* mutants have flowers with altered dorsal petals (Luo et al. 1996).

Reverse transcriptase polymerase chain reaction (RT-PCR) allows the amplification of gene transcripts (Figure 1.13(A)).

In situ hybridization is a technique for identifying specific mRNA transcripts within individual cells in tissue sections (Figure 1.13(B)).

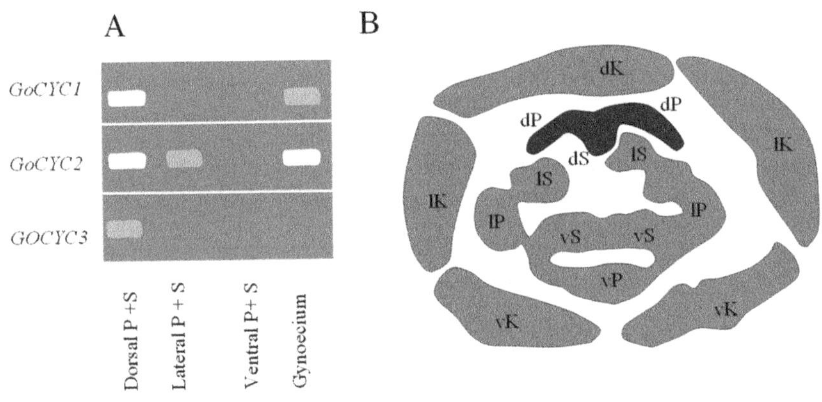

Figure 1.13. *(A) Reverse transcriptase polymerase chain reaction (RT-PCR) analysis of CYC-like genes in Gratiola officinalis (Plantaginaceae), from Preston et al. (2009). GoCYC1 and GoCYC2 expression is highest in the dorsal petals and stamens and gynoecium. GoCYC3 expression is relatively low in all tissues examined. (B) RNA in situ hybridization was used to detect the spatial expression of a CYC-like gene in a stage 6 flower (after carpel initiation) in Bournea leiophylla (Gesneriaceae): expression is detected in the dorsal petals and stamen (from Zhou et al. 2008). K = sepal; P = petal; S = stamen; d = dorsal; l = lateral; v = ventral*

Box 1.6. *RT-PCR and in situ hybridization methods for detecting gene transcripts*

In contrast to the peloric flowers of *A. majus*, which are ventralized, the *A. majus* backpetals mutant has dorsalized lateral and ventral petals. SEM analysis showed that flower development in the backpetals mutant is indistinguishable from that of the wild type, up to stage 6, when all the floral organs have been initiated. However, by stage 7, the backpetals mutant could be distinguished from the wild type by the larger size and rounder shape of the lateral petals. By stage 8, the lobes of the lateral and ventral petals in the backpetals mutant resemble the dorsal lobes, although with a more symmetrical shape (Luo et al. 1999).

In situ hybridization failed to detect any abnormal expression of either *CYC* or *DICH* in the early stages of floral development of backpetals mutant (stage 4). However, by stage 6, *CYC* expression was detected in the lateral petals. RT–PCR analysis of cDNA from dissected petals from more mature flower buds showed that *CYC* expression had expanded into the lateral and ventral petals, while the expression pattern of *DICH* was unchanged (Figure 1.14) (Luo et al. 1999).

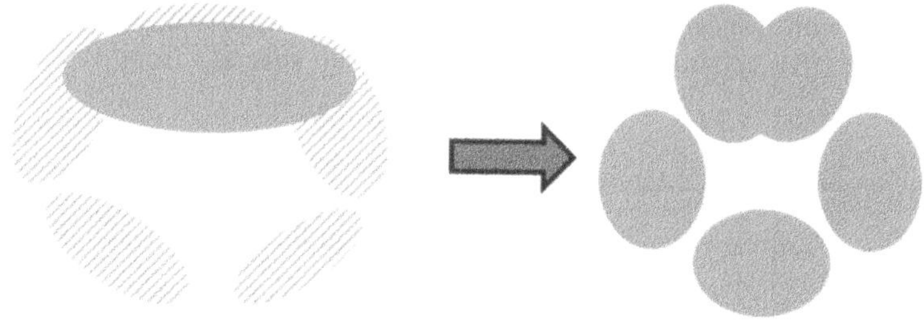

Figure 1.14. *CYC expression (in blue) in the backpetals mutant of A. majus at stages 4 (left) and 6 (right) (from Luo et al. 1999)*

Horn et al. (2015) conducted a comprehensive search of published land plant genome and EST databases using *TCP18*, *TCP1* and *TCP12* from *Arabidopsis thaliana* as representatives of the core eudicot specific *CYC1*, *CYC2* and *CYC3* clades, respectively. No *CYC-like* genes could be detected in the sequenced genomes of the bryophyte *Physcomitrella patens*, the lycophyte *Selaginella moellendorfii* nor in the gymnosperms *Pinus taeda* and *Picea abies*. However, in the sequenced genome of the earliest diverging angiosperm, *Amborella trichopoda*, one *TCP* gene with similarity to the *CYC-like* clade was identified. EST dataset searches detected *CYC*-like *TCP* genes in *Nuphar advena* (Nymphaeales), *Liriodendron tulipifera* (Magnoliales) and *Persea americana* (Laurales). As no *CYC-like* genes could be identified from available bryophyte, lycophyte and gymnosperm genome databases, it appears that members of the *CYC-like* clade first evolved in the common ancestor of angiosperms.

Box 1.7. *The CYC-like clade first evolved in the common ancestor of angiosperms*

1.5. Curie's symmetry principle

Curie's symmetry principle (1894) states that "when certain causes produce certain effects, the elements of symmetry of the causes must be found in the effects produced". With this definition, Pierre Curie extended the application of the symmetry principle used by the crystallographer Auguste Bravais (1850) – who, with his brother Louis Bravais, was also a pioneer in the study of phyllotaxis (1837) – to all physical phenomena.

Here, we propose to apply Curie's principle to the genetic causes of symmetry in angiosperms. The cause would be the symmetry of the expression of *CYC*-like genes on the corolla, and the effect would be the symmetry of this corolla.

In *Antirrhinum* and its mutants, the absence of expression of these genes (in the peloric mutant) or expression on all the petals (in the backpetals mutant) corresponds to actinomorphic corollas (Figures 1.12(D) and 1.14). An asymmetrical expression with respect to the transverse axis corresponds to a zygomorphic corolla (Figure 1.12(B)). Can we apply this rule to all species, from Ranunculales to Dipsacales?

There would then be four possible cases:

– *Lack of expression of CYC-like genes*: the corolla is radially symmetrical.

– *Expression of CYC-like genes throughout the corolla*: the corolla is radially symmetrical.

– *Asymmetrical expression of CYC-like genes with respect to the transverse axis*: the corolla is symmetrical with respect to the dorsoventral axis and asymmetrical with respect to the transverse axis (zygomorphy or "asymmetry along the dorsoventral axis", according to biologist words).

– *Disymmetrical expression of CYC-like genes*: the corolla is disymmetric.

2

Spirals Are Symmetrical Figures

2.1. The model

The growth of a meristem can be characterized by the radius R_0 of its central zone, the centrifugal velocity V of the lateral organs moving away from the center as the plant grows and the time T between the appearance of two consecutive primordia (the plastochron). VT is the distance traveled by a primordium away from the central zone during one plastochron. The ratio of this distance to the radius of the central zone R_0 gives the growth index $G = VT/R_0$, a dimensionless number that characterizes in a unique manner the growth of the meristem, irrespective of size and time scale (Godin et al. 2020). For instance, on day 11, the growth index of *L. usitatissimum* shoot apical meristem (SAM) is 0.08 (Figure 1.5(A)) (Walch and Blaise 2023).

Here, we follow the Douady and Couder model (1992, 1996a) and describe the initiation of primordia as depending on the intensity of an inhibition potential at the meristem front. The meristem front is represented by a circle of fixed radius $R_0 = 1$ chosen as the unit of length, and primordia are points located between this front and the outer edge of the meristem. The initiation of a new primordium (i) takes place on the front at the point of polar coordinates (R_0, θ_0), where θ_0 is the initiation angle that minimizes the inhibition potential Ui generated at the front by all preexisting primordia. This potential decreases exponentially with the distance from the primordium generating the inhibition field, the decay length being γ.

For a color version of all the figures in this chapter, see www.iste.co.uk/walch/phyllotaxis2.zip.

If d_{ij} is the distance between the new primordium i and any preexisting primordium j (between 1 and i – 1) at a distance Rj from the center of the meristem and with an angular coordinate θ_j:

$$d_{ij} = \sqrt{R_0^2 + R_j^2 - 2R_0R_j\cos(\theta_i - \theta_j)} \qquad [2.1]$$

then the potential is given by:

$$U_i(\theta) = \sum_{j=1}^{i-1} \exp(-\frac{d_{ij}}{\gamma}) \qquad [2.2]$$

Scaling the cartesian coordinates by a constant c:

$x \to cx$; $y \to cy$ and $\gamma \to c\gamma$ that is $d_{ij} \to cd_{ij}$ and $\gamma \to c\gamma$

does not change the inhibition potential, that is, the model is scale invariant:

$U(\theta) \to U(\theta)$

The dimensionless quantity characterizing the range of the inhibition interaction is the "effective distance":

$d = R/\gamma$

which is the ratio of the actual distance (R) divided by the decay length of the interaction (γ). The model thus produces the same pattern in plants with the same effective distance, irrespective of the real diameter of their meristem.

Box 2.1. *Definition of effective distance*

Our model differs from the models of Douady and Couder (1996b), Smith et al. (2006) and Kitazawa and Fujimoto (2015), which all assume that a constant threshold potential value determines primordia initiation and plastochron ratios. Here, the model parameters are not the threshold potential value but the plastochron ratios (r_i) or the distance from the center of meristem (R_i) (see Appendix 1 in section 2.6.1):

$$r_i = \frac{R_i}{R_{i+1}} = \exp(G_i) > 1 \qquad [2.3]$$

Since organs are in contact, high plastochron ratios are equivalent to organs being large, whereas small plastochron ratios (closer to 1) are equivalent to organs being small (Figure 1.6).

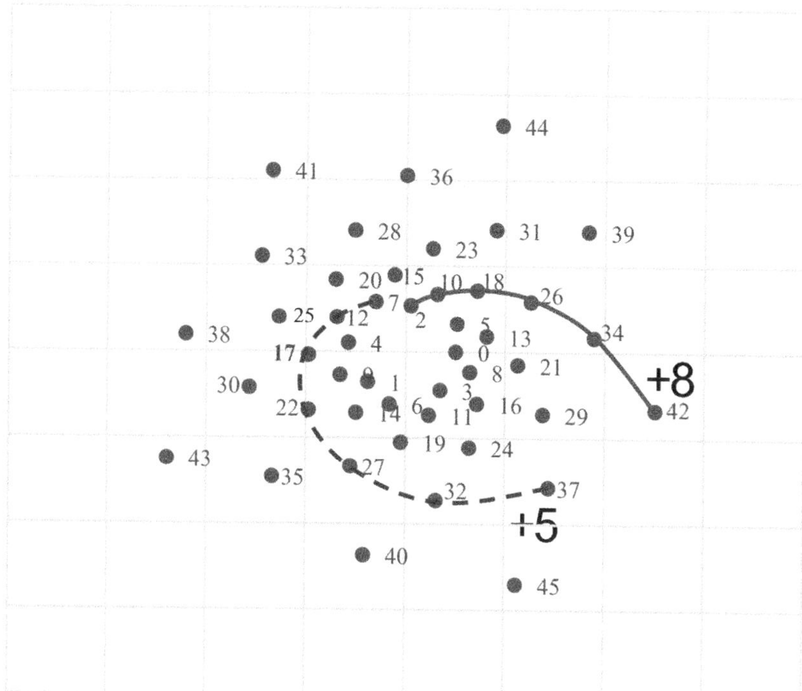

Figure 2.1. *Centric lattice of primordia according to the model for $\phi = 0.382$ (the golden angle) and G = 0.04 (corresponding to the (5, 8) phyllotaxis mode). Spirals are +5 and +8 parastichies. Numbers correspond to primordia from the youngest to the oldest*

Our model equations [2.1]–[2.3] allows us to calculate the position of all primordia for given values of the divergence angle and the growth index (Figure 2.1).

2.2. Phyllotaxic spirals are symmetrical figures

We define a steady state such that the divergence angle (ϕ) and the growth index (G) are constant. In a cylindrical representation (a Bravais lattice), the points in space (h, θ) are therefore aligned. The line segments thus formed are the parastichies (Figure 2.2).

$$h_{k+1} = h_k + G \quad [2.4]$$

$$\theta_{k+1} = \theta_k + \phi \quad [2.5]$$

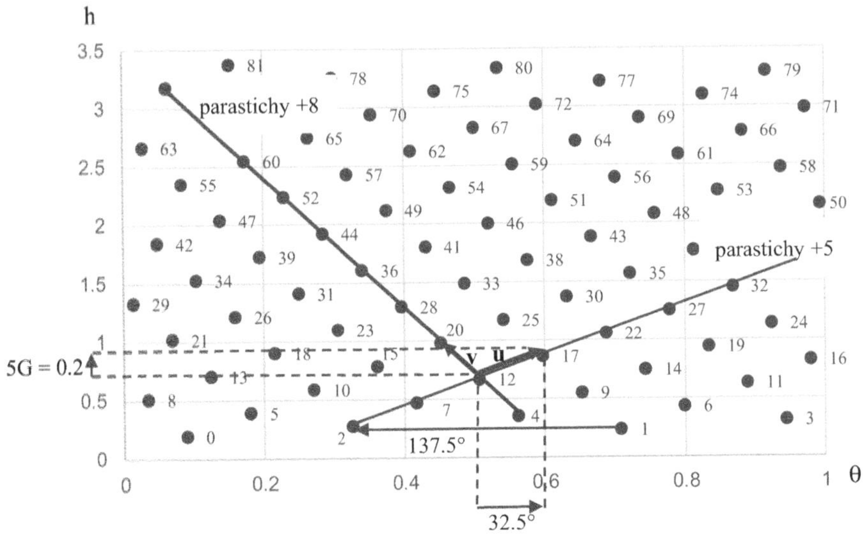

Figure 2.2. *Cylindrical lattice representing the position of 82 primordia for $\phi = -0.382$ (the clockwise golden angle) and G = 0.04 (phyllotaxis mode (5, 8))*

The correspondence between cylindrical and centric representations was established by Rothen and Koch (1989) (see Appendix 2 in section 2.6.2).

A system is symmetric if a transformation applied to all points of the system displaces the initial point to a place occupied by another point of the system. Thus, the corolla of *Lamium galeobdolon* presents axial symmetry (zygomorphy), its right side being transformed into the left side (Figure 2.3): with physicists we say that the corolla of *L. galeobdolon* (Lamiaceae) is invariant by axial symmetry.

A translation by vector **u** or **v** leaves the cylindrical lattice in Figure 2.2 invariant: all points of the transformed lattice are superimposed on points of the original lattice (e.g. 12 → 17).

On the centric figure, this translation corresponds to a rotation of $-5 \times 137.5°$ $= -327.5° = 32.5°$ and a dilatation of ratio 1.04^5 (see Appendix 2 in section 2.6.2). Thus, point 12 in Figure 2.4 is in the same place as point 17 in Figure 2.1.

Thus, for a constant divergence angle and growth index, phyllotaxis is invariant under two transformations corresponding to the two phyllotaxis modes (e.g. +5 and +8 on clockwise and counterclockwise parastichies). Phyllotaxic spirals symmetrical figures.

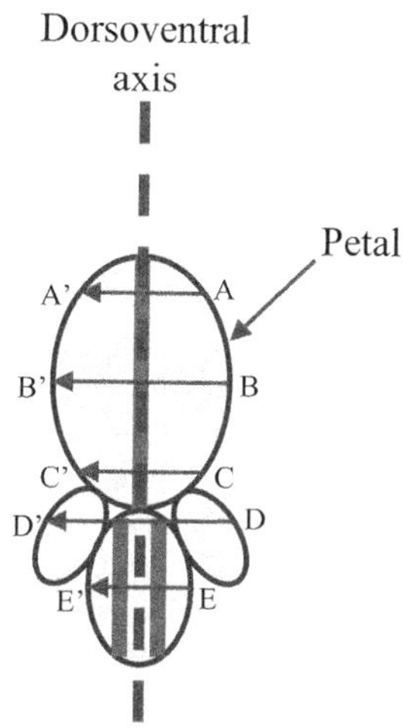

Figure 2.3. *Axial symmetry of the corolla of Lamium galeobdolon*

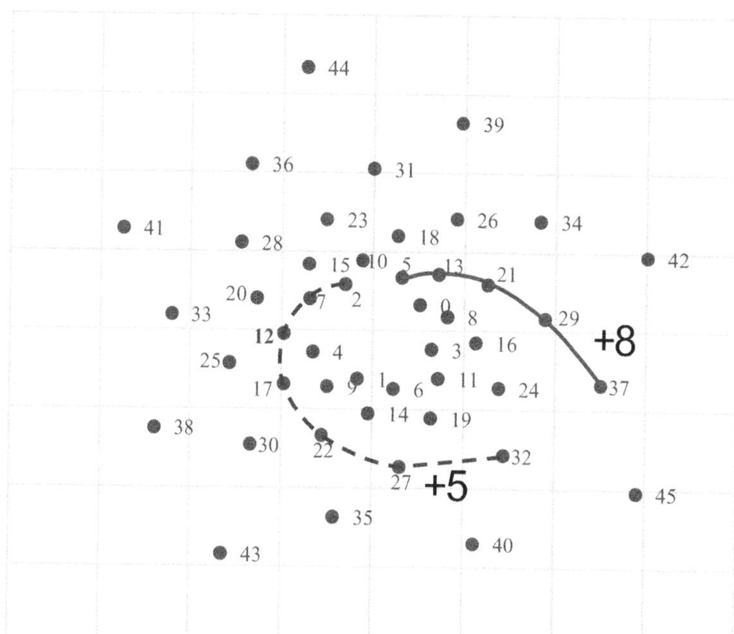

Figure 2.4. *Transformation of the centric representation in Figure 2.1 by a rotation of 32.5° and a dilatation of ratio 1.04^5*

2.3. Projection of the three-dimensional meristem

Williams (1975) showed a three-dimensional drawing of the SAM of *Linum usitatissimum,* the apex being surrounded by leaf primordia and young leaves measuring up to 1.5 mm compared to meristem diameter of 0.35 mm. Figure 2.5(A) shows a paraboloid form of this meristem with attachment areas for all of the primordia on the near side.

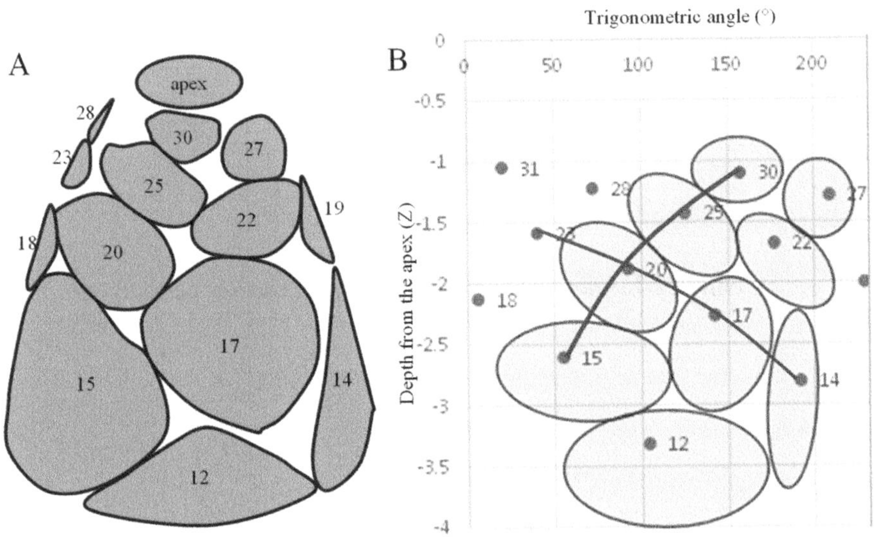

Figure 2.5. *(A) Conic projection of the three-dimensional shoot apical meristem of Linum usitatissimum, according to Williams (1975). (B) Model of cylindric projection with parastichies. Numbers correspond to the chronological order of leaf primordia initiation. Phyllotaxis mode is (3, 5)*

According to Richards (1951), the plastochron ratio as measured on the transverse sections of a cone (distance (X_i) from O of primordium i transverse projection over the distance of its chronological successor X_{i+1}) is given as (Figure 2.6(A)):

$$r = \frac{OX_i}{OX_{i+1}} = \frac{AZ_i}{AZ_{i+1}} \qquad [2.6]$$

where Zi is the vertical distance of primordium i to the apex.

Therefore, the height of the primordia (Z) is proportional to their distance from the apex on a transverse projection (X). The measurements of Williams (1975) (Table 2.1) allow us to model the cylindrical projection of the cauline meristem of *Linum usitatissimum* (corresponding to the centric lattice of Figure 2.1) taking into account the evolution over time of plastochron ratios (Figures 2.5(B) and 2.6(C)).

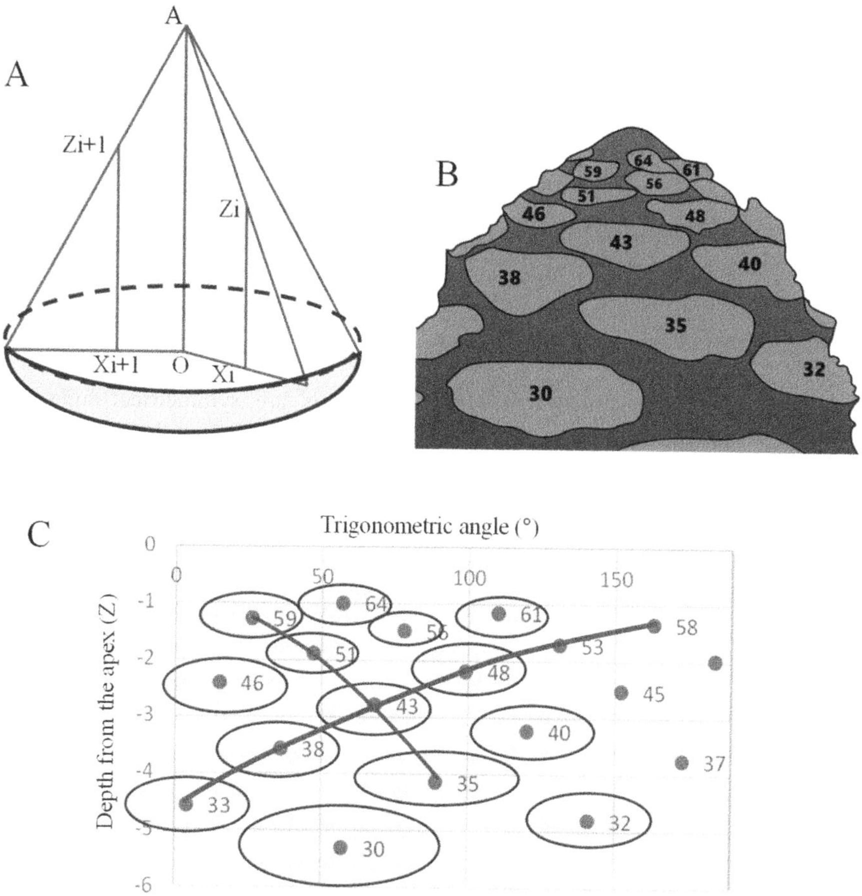

Figure 2.6. *(A) Diagram illustrating the relation between the conical presentation of a meristem and its projection on a transverse plane according to Richards (1951). (B) Conic projection of the three-dimensional cauline meristem of Linum usitatissimum (redrawn from Meicenheimer 2006). (C) Model of cylindrical projection. Numbers correspond to the chronological order of leaf primordia initiation*

Age (days)	Primordium number	Plastochron ratio
1	Cot.–2	1.330
4	2–4	1.189
4	4–6	1.152
8	4–14	1.101
11	10–20	1.082
15	20–30	1.073
18	35–45	1.052
22	51–61	1.049

Cot. = cotyledons

Table 2.1. *Evolution of the plastochron ratio with primordium age in Linum usitatissimum (from Williams 1975)*

2.4. Examples of spiral symmetry

We have demonstrated that Fibonacci spirals are symmetrical figures: this explains why we can observe similar spirals in worlds other than that of plants.

2.4.1. *Plant lattices*

Lepidodendron (from the Greek λεπίς scale, and δένδρον, tree) is an extinct genus of primitive vascular plants known as scale trees or arborescent lycophytes. They thrived during the Carboniferous period (360–300 million years ago) and are part of the coal forest flora. They grew up to 50 m in height and 1 m in diameter. Fossilized scales at the base of their leaves on the stem show spiral patterns (Figure 2.7).

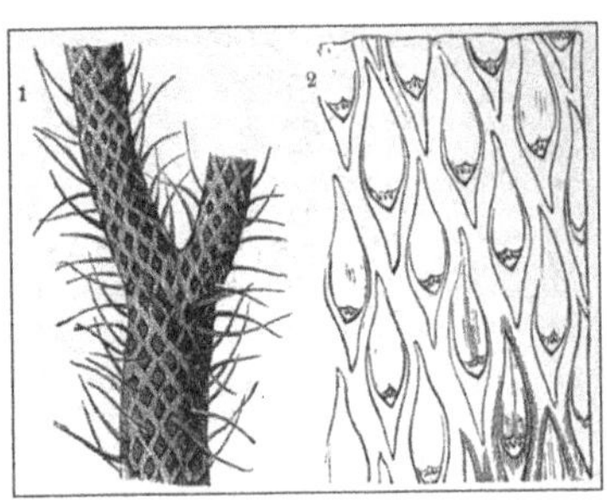

Figure 2.7. *Drawing of a Lepidodendron stem from The American Cyclopedia (1879)*

The fossil remains of *Lepidodendron* from coalfields in the Glasgow area were studied by Dickson (1871). He determined leaf phyllotaxis in 13 specimens with more or less cylindrical stems. Three of these presented Fibonacci spirals (i.e. (8, 13), (13, 21), (55, 89) modes), four had multijugate Fibonacci phyllotaxis (e.g. (15, 24)), one showed Lucas phyllotaxis (18, 29) and the rest had (9, 14), (12, 19), (10, 13) and (23, 35) phyllotaxis modes. According to Dickson, the phyllotaxis of *Lepidodendron* is more variable than that of living plants such as *Cactus*.

In angiosperms, the phyllotaxis of leaves is established in the SAMs (e.g. *Linum usitatissimum*) (Figure 2.5 and 2.6).

The inflorescence of snapdragon, *Antirrhinum majus* (Plantaginaceae), is a simple raceme with an apical meristem that produces floral meristems following a (3, 5) Fibonacci phyllotaxis (Figure 1.10).

The floral apex of *Magnolia soulangeana* shows a quite regular Fibonacci pattern with a (5, 8) mode for androecium (Zagórska-Marek 1994) (Figure 2.8).

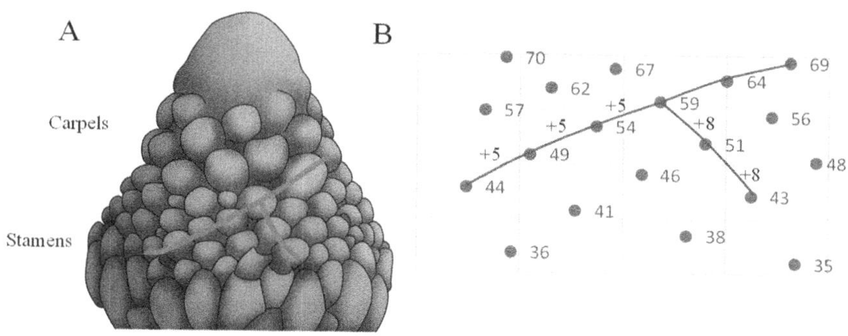

Figure 2.8. *(A) The floral apex of Magnolia soulangeana during stamen and carpel initiation, redrawn from Zagórska-Marek (1994). Two partial parastichies are shown. (B) Model of a cylindrical representation of floral organ initiation in M. soulangeana (Walch and Blaise 2023). The same parastichies as in (A) are shown*

2.4.2. *Disc-stacking in a cylinder*

In 2020, Godin, Golé and Douady published "Phyllotaxis as geometric canalization during plant development". They model phyllotaxis by stacking discs one by one on the surface of a cylinder in the lowest possible place above the previous discs, without overlap (Figure 2.9). Their approach gives fundamentally equivalent results to ours in terms of primordia initiation at the front, rising phyllotaxis to noble angles (the van Iterson diagram or our phyllotaxic frieze), organ permutations and quasi-symmetric arrangements (Golé et al. 2016; Godin et al. 2020; our results in Walch and Blaise 2023, Chapter 6).

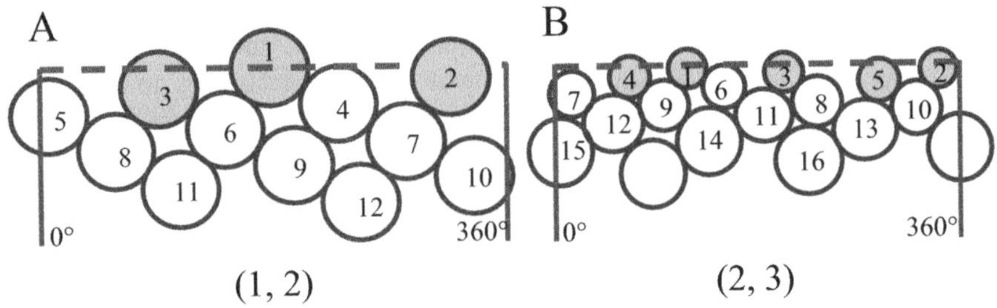

Figure 2.9. *Cylindrical representation of (A) trimerous and (B) pentamerous fronts (dashed lines) according to phyllotaxis mode from Godin et al. (2020)*

However, the physical basis of the two models is very different. The disc-stacking model relies on contact pressure between primordia, represented as incompressible discs, and pressure exerted by a "wall" on these organs. The rise in phyllotaxis mode is determined by the rate of decrease of the organ size relative to the diameter of the meristem and by competition between primordia for space. Development is thus channeled in both the literal (by the walls) and figurative sense: phyllotaxis is primarily governed by geometric developmental constraints.

However, according to our model, the interaction between organs is not just a purely mechanical force exerted by hard discs in direct contact and there is no wall imprisoning the meristem. The inhibition field plays an important organizational role even when two primordia are distant from each other and organs are soft.

The results of the disc-stacking model are only equivalent to those of the molecular inhibition model because the disc-stacking model is also of spiral symmetry.

2.4.3. *Levitov's energy model*

Levitov (1991) modeled a regular lattice of interacting deformable objects that repulse each other (Lee and Levitov 2021). Lattice geometry is controlled by the balance between external compression and repulsive interactions between different lattice points. The repulsive interaction is determined solely by the distance D between interacting points. It can be, for example, exponential ($e^{-D/\gamma}$) or a power law ($D^{-\gamma}$).

Levitov demonstrated that this energy is invariant under the modular transformation SL(2, Z):

$$z \to z' = \frac{az+b}{cz+d} \qquad [2.7]$$

with

$$\det\begin{pmatrix} a & b \\ c & d \end{pmatrix} = 1 \qquad [2.8]$$

In comparison to the model of Godin et al. (2020), objects are deformable, but still subjected to stress arising because of closed-volume or confinement constraints (Levitov nevertheless wonders what the origin of such a stress in plant growth can be). The system deforms in such a way to minimize the total repulsion energy for a given stress value. With an increase of the external stress, the lattice structure climbs the Fibonacci sequence.

2.4.4. *The Douady–Couder experiment*

In the Douady–Couder experiment (Douady and Couder 1996a), a horizontal teflon dish filled with silicone oil is placed in a vertical magnetic field (Figure 2.10). This field has a weak radial gradient: it is minimal at the center and maximal at the periphery of the dish. Organ primordia are simulated by falling drops of ferrofluid: in a magnetic field, they are polarized and form dipoles that are attracted toward the region of maximum field. The drops thus spontaneously drift from the center to the border of the dish and repel each other. The repulsion caused by previous drops forces a new drop to place itself in the largest available space at the periphery of the central region.

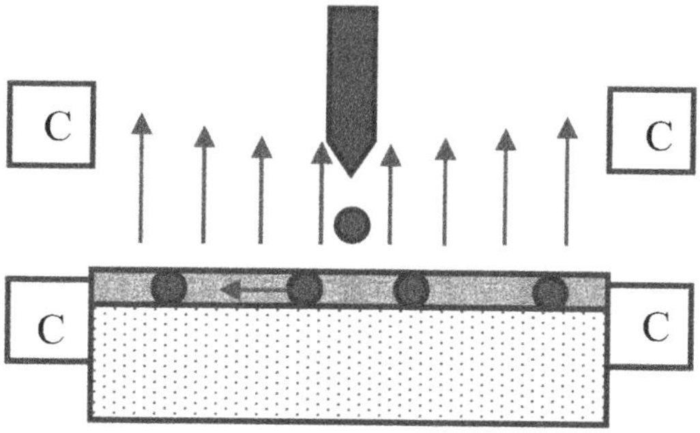

Figure 2.10. *The device used in the Douady–Couder experiment. C = two Helmholtz coils producing a magnetic field. Vertical arrows = magnetic field. Discs = drops of ferrofluid. Horizontal arrow = centrifugal movement of the drops*

As these drops form identical parallel dipoles, they repel each other with a force that is proportional to D^{-4} (D being the distance between drops). This system gives rise to self-organizing spiral structures. The observed patterns (e.g. Fibonacci spirals) depend on the periodicity of drop deposition (T), the radius of the circle outside of which the angular position of the particles does not change (R_0) and the advection velocity V controlled by the magnetic field gradient.

2.4.5. *The magnetic cactus*

The "magnetic cactus" consists of equally spaced permanent magnets mounted along an axis around which they are free to rotate (Nisoli et al. 2009). All magnets are outward facing to produce a repulsive interaction between magnet pairs (Figure 2.11). At the beginning of each experiment, the magnets are positioned in a random way. They then position themselves and, in the final state, give rise to spiral structures with angles of divergence between two successive magnets that are equal to 137.5°, 99.5° or even 77.96° depending on the length of the magnets.

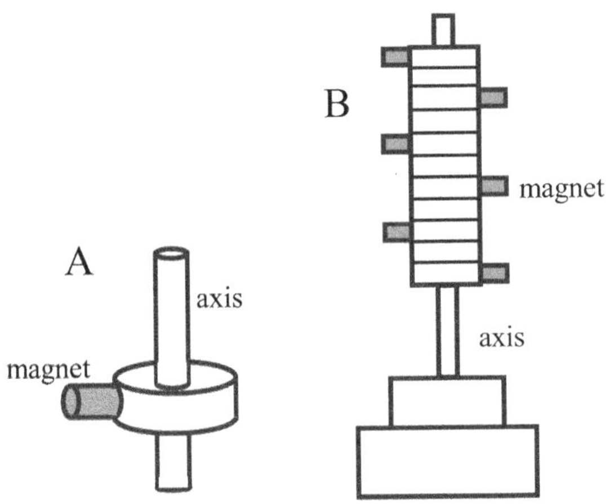

Figure 2.11. *The experimental magnetic cactus apparatus. (A) Each unit of the magnetic cactus consists of a magnet element and a ring secured to the central axis. (B) Schematic representation of the mounted magnetic cactus. From Nisoli et al. (2009)*

Nisoli et al. (2009) used a magnetic dipole–dipole interaction and showed that the lattice energy was minimal for the three so-called (Jean 1994) "noble" divergence angles.

2.4.6. *Contact pressure*

In 1878, the Swiss botanist Simon Schwendener proposed a contact pressure model of phyllotaxis, namely that the arrangement of leaves is the consequence of the pressure exerted by each primordium on neighboring primordia.

For Schwendener, the two primordia located on the two parastichies crossing a given primordium (i) produce, by contact, a pressure on this primordium (i). Using geometric methods that were inspired by those used in architecture for gabled roofs formed by sloping rafters of unequal lengths, he believed he had shown that starting from a (1, 2) mode, phyllotaxis converges toward the golden angle. However, his proof was wrong (Adler et al. 1977).

For Irving Adler et al. (1977), contact pressure is represented by the hypothesis that the minimum distance between lattice points is maximized. They showed that if (m, n), with m < n, is the contact phyllotaxis mode resulting from when contact pressure first becomes effective, then continued contact pressure leads to higher phyllotaxis modes as the plastochron ratio decreases and must follow consecutive terms of the Fibonacci sequence generated by the numbers m and n, namely, m, n, m + n, m + 2n, 2m + 3n and so on.

Ridley (1982) modeled phyllotaxis as the consequence of a two-step biological process: first primordia initiation, followed by their interaction during meristem growth. He showed that the divergence angle converged rapidly toward the golden angle, irrespective of the value of the divergence angle between the first primordia. After the initiation of a primordium, the contact pressure exerted by all existing primordia was calculated and produced a rearrangement of these primordia. Primordia movement under the effect of contact pressure was calculated for each primordia pair (provided they were close enough), starting from the youngest to the oldest primordium, as follows: primordium (i) moves away from primordium (j) by a vector whose direction is that of the axis (i → j) and by a length that is less than the distance dij. Several iterations of this algorithm were performed before allowing the initiation of a new primordium.

According to Hellwig et al. (2006) models of the effect of contact pressure, provided primordia are initiated between 127° and 149°, moves primordia to a position where they are as far apart as possible from each other, and creates Fibonacci spirals of high modes. Even better results were obtained when new primordia were initiated according to Hofmeister's rule.

2.5. Phyllotaxis as a dynamical system

2.5.1. *The dynamics of phyllotaxis*

We calculated the values of the inhibition potential (U) as a function of the divergence angle (ϕ) and the plastochron ratio (r) using equations [2.1]–[2.3] (Walch 2022).

In dicots, stem phyllotaxis begins with two opposite facing cotyledons (ϕ = 180°). This state is stable (the potential is minimal) until the plastochron ratio reaches a critical value (r ≈ 2.2) from which two minima arise corresponding to the two opposite chiralities (clockwise and counterclockwise). Phyllotaxis becomes spiral and its mode is (1, 2) (Figure 2.12). This type of bifurcation of a potential is well known in physics under the name of Hopf bifurcation, which applies to second-order phase transitions with a change in the symmetry of the system, for example, in the Rayleigh–Bénard convection. Here, it marks the transition from a distichous phyllotaxis to a spiral phyllotaxis (spiral symmetry). The two chiralities are equiprobable.

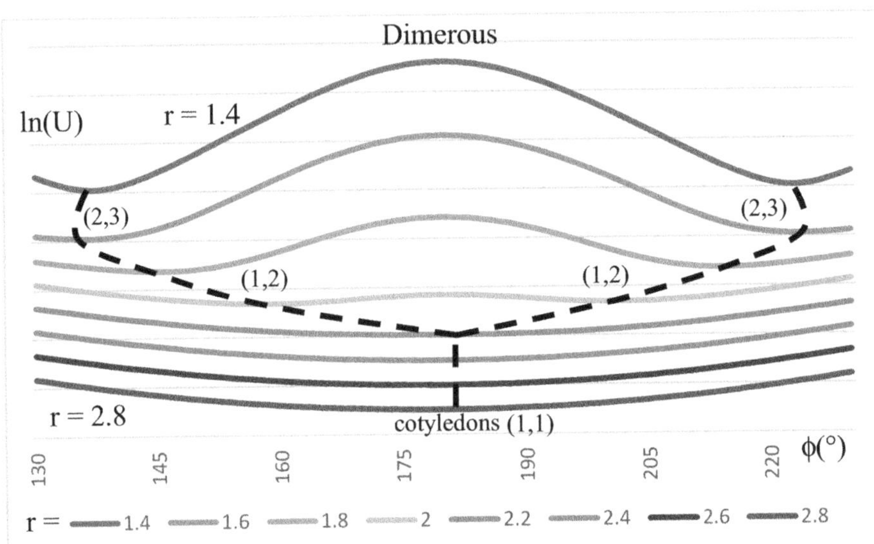

Figure 2.12. *The inhibition potential (ln(U)) as a function of the divergence angle (ϕ) for different plastochron ratios (r) between 2.8 and 1.4 as calculated from the model. For each value of r, the curve of the natural logarithm of the inhibition potential is shown; trajectories in phase space as r decreases climb the "valleys" (local minima) formed by the inhibition potential curves (dashed lines)*

Primordia initiate along the "valleys" (local minima) of the inhibition potential. The dynamics of the system as the plastochron ratio decreases is consistent with that of van Iterson's diagram (Douady and Couder 1996a): the same bifurcations are found, the number of inhibition potential minima increasing with a decrease in the value of the plastochron ratio (Figure 2.13). The system rises in the Fibonacci sequence by a succession of "quasi-bifurcations". Indeed, at each evolution, such as from (3, 5) to (5, 8), the passage to (3, 8) is impossible because there is a potential barrier (a "mountain") between the two modes.

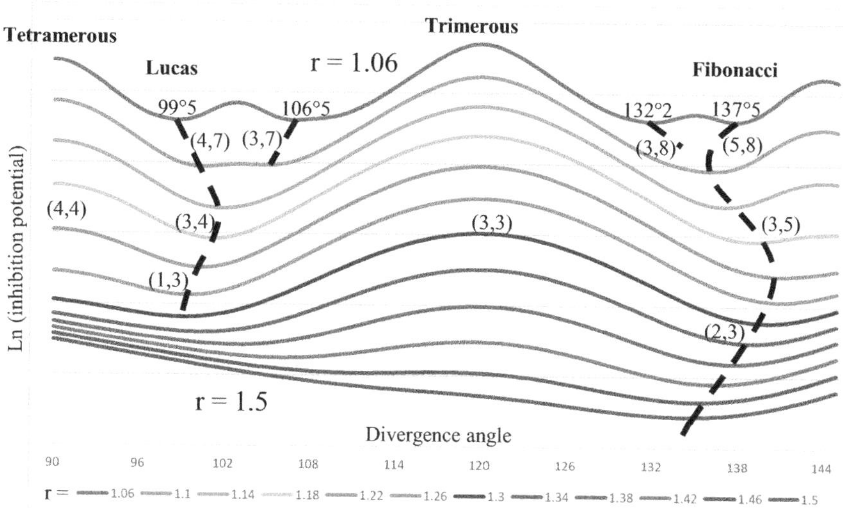

Figure 2.13. *The inhibition potential (ln(U)) calculated from the model as a function of the divergence angle (ϕ) for different plastochron ratios (r) between 1.5 and 1.06. For each value of r, the curve of the natural logarithm of the inhibition potential is shown; trajectories in phase space as r decreases climb the "valleys" formed by the inhibition potential curves (dashed lines = local minima). The angles given above the curves are the noble angles toward which each trajectory converges. Note that (4, 4) and (3, 3) are whorled tetramerous and trimerous phyllotaxis ($\phi = 90°$ and $\phi = 120°$), respectively*

In the vast majority of plants with spiral phyllotaxis, the number of parastichies follows the terms of the Fibonacci sequence. Some plants, however, are arranged according to other sequences, and the most common of these is the Lucas sequence (2, 1, 3, 4, 7, 11, etc.).

Only Fibonacci phyllotaxis can develop in continuity with the first two cotyledons. Indeed, when plastochron ratios decrease, only quasi-bifurcations appear, the minimum potentials of the branches leading to angles of divergence that are different from the golden angle appearing in discontinuity with the parent branch.

This is the reason for the very high frequency of Fibonacci phyllotaxis compared to other types of phyllotaxis. "Accidents" must happen during development for other types of spirals to occur in a plant, the most frequent being Lucas phyllotaxis, the first to appear (of mode (1, 3)) for a plastochron ratio $r \approx 1.3$ (see section 2.5.7).

2.5.2. *Whorled phyllotaxis*

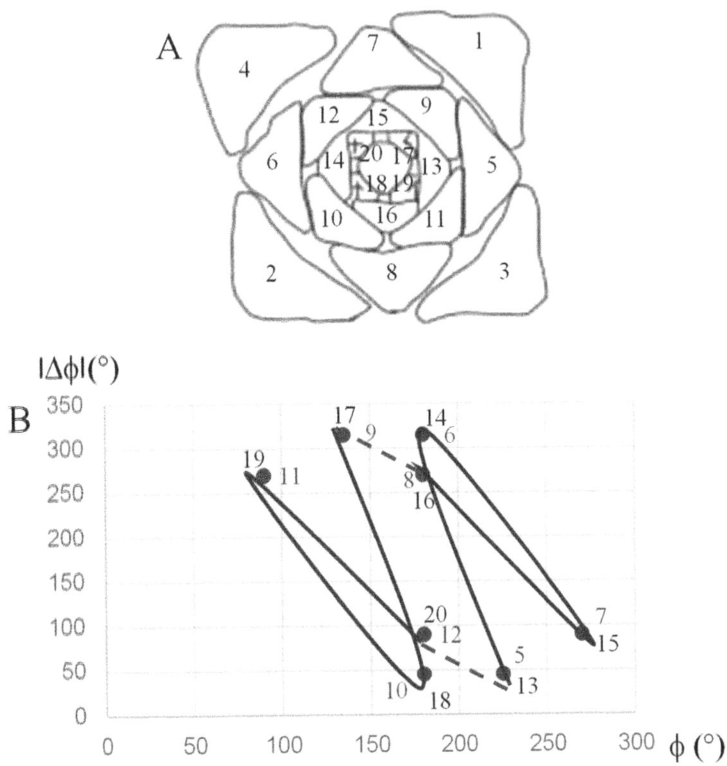

Figure 2.14. *(A) Cross-section of the apical vegetative meristem of Plantago webbii (Plantaginaceae), from Rutishauser (1998). (B) Phase portrait of Plantago webbii*

Spiral arrangements are not the only type of phyllotaxis. Whorled patterns consist of alternating whorls, each composed of the same number of organs (merosity). Divergence angles between organs within a whorl are constant and equal to 360° divided by the number of organs: for instance, the divergence angle is = 360°/2 = 180° in a dimerous whorl, 360°/3 = 120° in a trimerous whorl and 360°/4 = 90° in a tetramerous whorl. Whorled patterns present perfectly vertical orthostichies unlike the inclined pseudo-orthostichies of spiral phyllotaxes.

The vegetative meristem of *Plantago webbii* (Plantaginaceae, Lamiales) is tetramerous. The four members of one whorl are initiated simultaneously at the shoot apex (Figure 2.14(A)) (Rutishauser 1998). Each whorl has the same number of primordia.

In dynamical system theory, the phase portrait represents the evolution of a system in phase space, here (ϕ, $\Delta\phi$), where ϕ is the divergence angle between points i + 1 and i and $\Delta\phi$ is the difference between the divergence angles at point i + 1 and point i (we have chosen abs($\Delta\phi$) for more visibility).

The whorled vegetative meristem of *Plantago webbii* walks on a "butterfly" of eight points and comes back to each point at each orthostichy (Figure 2.14(B)). In physics, the archetype of a periodic phase portrait is a simple pendulum without friction. The dashed lines correspond to the transition from a Greek cross to a Saint Andrew's cross and reciprocally.

2.5.3. *Fibonacci spiral phyllotaxis*

The plastochron ratios measured in the vegetative meristem of *L. usitatissimum* (Williams 1975), with Fibonacci spiral phyllotaxis (Table 2.1), can be approximated by a decreasing exponential function of organ number (n):

$$r(n) = r_0 e^{-\rho n} + r_{asympt} \qquad [2.9]$$

with r_0 = 0.35 (the initial value), ρ = 0.18 (the decrease rate) and r_{asympt} = 1.05 (the level of the asymptote) (Walch and Blaise 2023, Chapter 6). With these parameters and starting with two opposite primordia (cotyledons), we obtained a rapid convergence of the divergence angle toward the golden angle and a (5, 8) phyllotaxis mode. We call these parameters "the canonical parameters": Fibonacci spiral phyllotaxis requires a "slow and controlled" decrease in plastochron ratios.

We can compare the result of the model (Figure 2.15(B)) with Figure 2.15(A) (redrawn from Williams 1975) and appreciate how well the model fits the observations of this vegetative meristem.

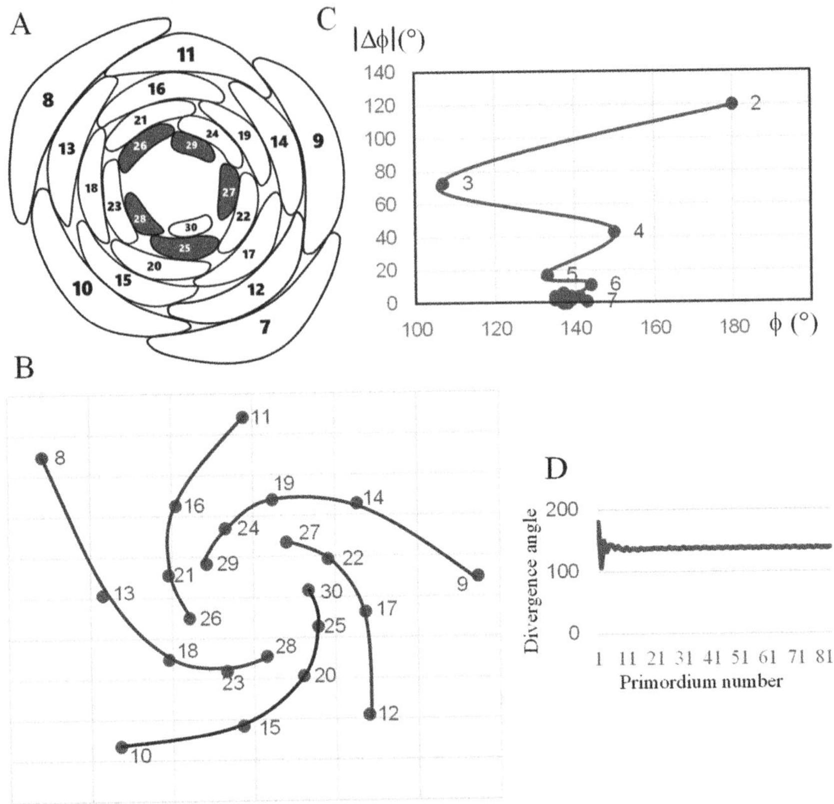

Figure 2.15. *(A) Representation of a 15-day old vegetative meristem of Linum usitatissimum from Williams (1975). (B) Model of the vegetative meristem of L. usitatissimum at day 15 of development using the canonical parameters (r_0 = 0.35, ρ = 0.18 and r_{asympt} = 1.05). (C) Phase portrait of Fibonacci phyllotaxis. Numbers correspond to the chronological order of leaves initiated along the shoot. (D) Value of the divergence angle between successive primordia derived from the model with the canonical parameters for 81 primordia*

The trajectory of the phase portrait of the Fibonacci vegetative meristem of *L. usitatissimum* converges rapidly to the golden angle. In physics, the archetype of a convergent phase portrait is that of a pendulum with friction whose oscillations are damped. The golden angle is called a fixed point or attractor toward which the trajectory of the system converges.

2.5.4. *Periodic divergence angles*

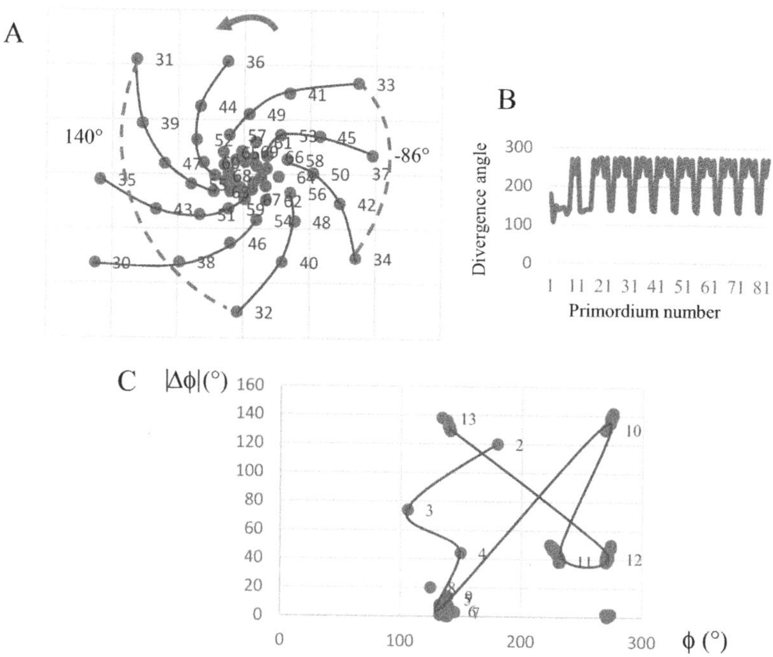

Figure 2.16. *The pattern obtained with r_0 = 0.35, ρ = 0.3 and r_{asympt} = 1.05: (A) centric representation of organ arrangement. (B) Value of the divergence angles showing a periodic pattern. (C) Phase portrait of periodic phyllotaxis*

If plastochron ratios decrease faster than the canonical value (ρ = 0.3 vs. ρ = 0.18), we observe a permutation of organs: for instance, the divergence angle between points 31 and 32 is 140° (counterclockwise), a value that is close to the golden angle, but that between points 33 and 34 is 360° – 86° = 274° (counterclockwise). Organ number 34 should be located +137° from point 33, however this is where point 35 is located, point 34 jumping instead to the next location and leaving its place to point 35 (Figure 2.16(A)).

As primordia are permutated, divergence angles oscillate between ≈137°, 230° and 274° (Figure 2.16(B)) and there are several attractors in the phase portrait (Figure 2.16(C)). This type of permutation has been observed in *Doryphora aromatica* (Laurales) (Staedler and Endress 2009; Walch and Blaise 2023).

2.5.5. *Chaotic/quasi-symmetric phyllotaxis*

For $r_{asympt} = 1.02$ and $\rho = 0.7$, we find nine pseudo-parastichies in both directions (i.e. a (9, 9) mode) (Figure 2.17). The inclination of the parastichies in both directions is more or less equal. Thus, phyllotaxis is quasi-symmetric.

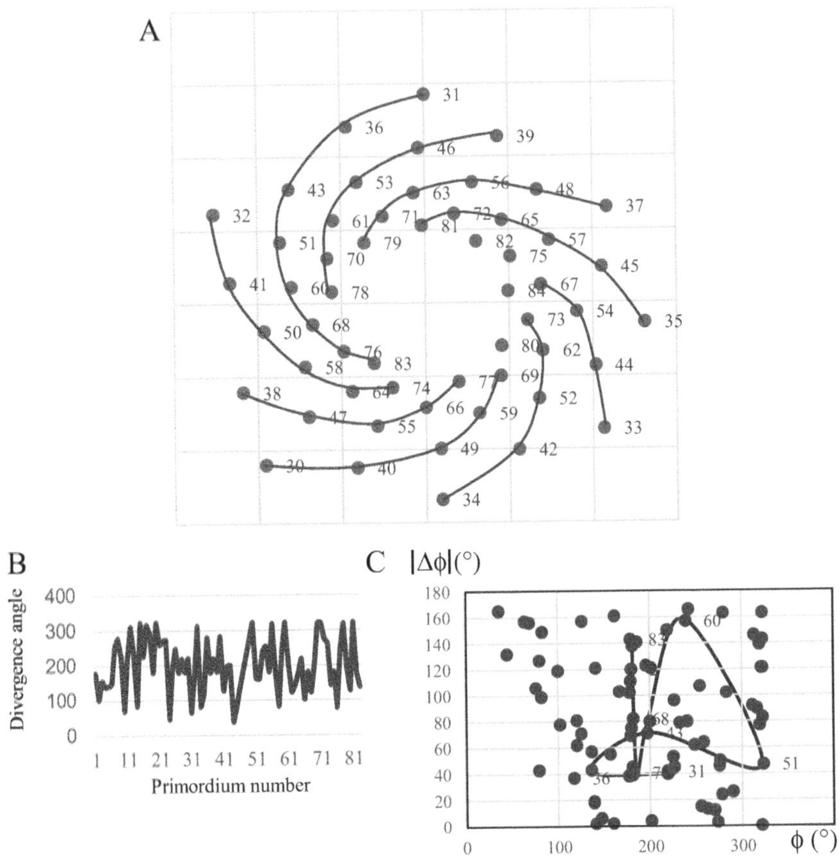

Figure 2.17. *Pattern obtained with $r_0 = 0.35$, $\rho = 0.7$ and $r_{asympt} = 1.02$: (A) Centric representation of organ arrangement. (B) Divergence angles showing a chaotic pattern. (C) Phase portrait of chaotic phyllotaxis*

Using the disc-stacking model, Golé et al. (2016) showed that a rapid decrease in disc diameter led to the breakdown of Fibonacci spirals in favor of a "quasi-symmetric" arrangement, with a nearly equal number of clockwise and counterclockwise spirals. This arrangement can be seen in inflorescences of

Spathypillium (Araceae), *Banksia* (Proteaceae), corn (*Zea mays*) and peteh (*Parkia speciosa*, Fabaceae subfamily Mimosoideae) (see Godin et al. 2020, Supplementary Information).

According to our model, if plastochron ratios decrease very rapidly, phyllotaxis becomes quasi-symmetric with a near equal number of left and right parastichies. This type of spiral phyllotaxis is different from Fibonacci phyllotaxis and can be mistaken for whorls. Pseudo-parastichies bring out an apparent order in the chaos of divergence angle values (Figure 2.17).

Floral phyllotaxis in *Nuphar* (Nymphaeaceae) has traditionally been thought to be spiral, a view supported by Eichler (1878); however, Endress and Doyle (2015) have asserted that all Nymphaeaceae have flowers with complex-whorled (i.e. non-spiral) phyllotaxis. The criterion used by Endress is as follows: phyllotaxis is whorled when the inclination of the left and right parastichies is the same. However, the criterion used by Endress and Doyle is insufficient: an equal inclination of left and right parastichies differentiates whorled phyllotaxis from Fibonacci phyllotaxis, but not whorled phyllotaxis from certain types of irregular spirals (Figure 2.18).

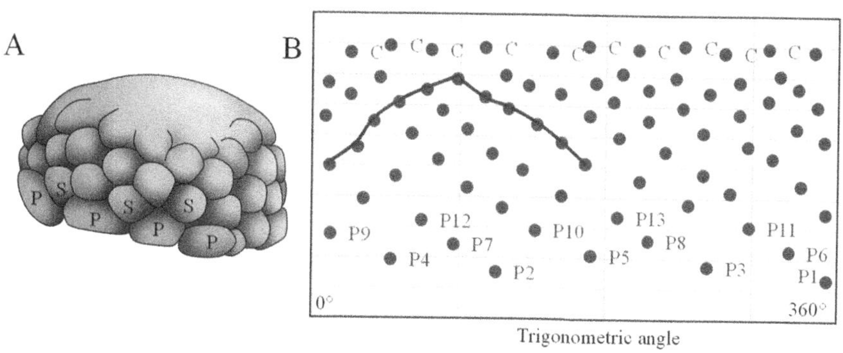

Figure 2.18. *(A) The floral meristem of Nuphar pumila (from El et al. 2020). (B) Cylindrical representation of the modeled floral meristem of Nuphar pumila showing quasi-symmetric pseudo-parastichies according to the model (from Walch and Blaise 2022a). P = petal (the associated number corresponds to the order of initiation); S = stamen; C = carpel*

According to Rutishauser (2016), the vegetative phyllotaxis of *Acacia* (Australian "wattles", Faboideae subfamily Mimosoideae) can be spiral (*A. amblygona* and *A. rossei*: Fibonacci system), "complex whorled" (*A. adoxa* with 9- to 10-merous phyllode whorls) or even irregular without any obvious order (*A. sphacelata subsp. verticillata* and *A. conferta*). The latter are correlated with geometrical parameters such as leaf and stamen primordia that are very small as compared to the size of the apical meristem.

According to Douady and Golé (2016), the male cone of *Cedrus* is a remarkable example of drift from Fibonacci to quasi-symmetric, as is the inflorescence of *Parkia speciosa*, which shows a solid quasi-symmetric pattern originating from a rapid transition.

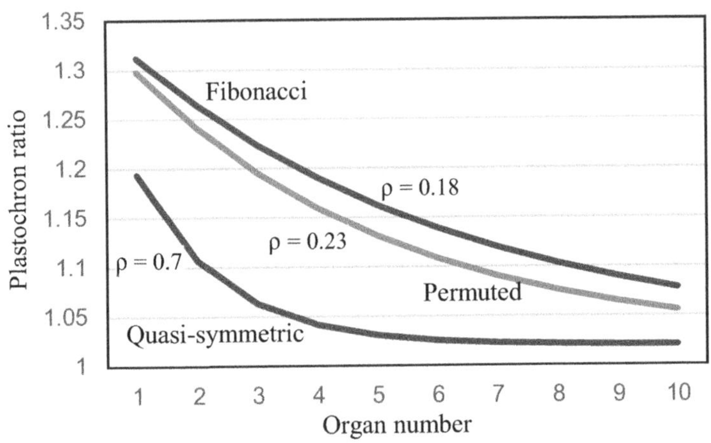

Figure 2.19. *Profile of plastochron ratios for three different decrease rates (γ = 0.2; r_0 = 0.35; r_{asympt} = 1.02)*

For $r_0 = 0.35$ and $r_{asympt} = 1.02$, the basin of attraction (the set of initial conditions leading to a long-time behavior that approaches an attractor) of Fibonacci phyllotaxis corresponds to values of the control parameter ρ less than 0.2, that of permuted phyllotaxis to values of ρ between 0.2 and 0.25, and that of quasi-symmetrical phyllotaxis to values of ρ greater than 0.25 (Figure 2.19).

Fluid flow behavior is controlled by similar dynamics. The transition from a laminar to a turbulent flow is determined by Reynolds number, which combines the velocity of the fluid, the width of the river in which it flows and fluid density and viscosity. When Reynolds number is small, as in a wide, gently sloping river, the flow is laminar. By contrast, a mountain torrent near its source (characterized by fast flowing water and a narrow width) will have a high Reynolds number: the flow is turbulent, and the movement of each drop of water is unpredictable. Between these two regimes, the flow is periodic.

2.5.6. *Pseudo-whorled phyllotaxis*

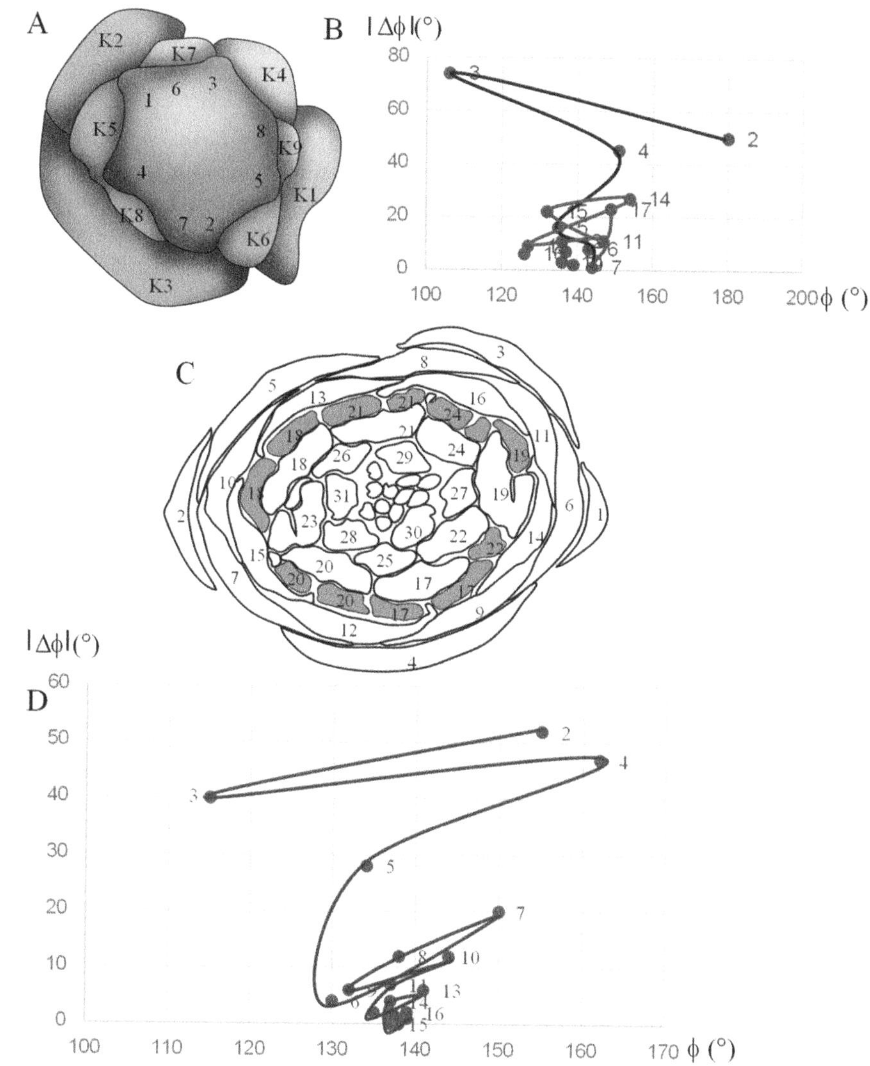

Figure 2.20. *(A) The floral meristem of Liriodendron tulipifera redrawn from an SEM micrograph from Erbar and Leins (1994) and (B) phase portrait of floral organ initiation in L. tulipifera, with r_0 = 0.35, ρ = 0.18 and r_{asympt} = 1.02; the plastochron ratio between pseudo-whorls is set to r = 1.30. (C) Transverse section of the flower bud of Daphnandra repandula, from Staedler and Endress (2009), showing organs belonging to distinct pseudo-whorls. (D) Phase portrait of D. repandula (R_0 = 0.5, r_0 = 0.18, ρ = 0.1 and r_{asympt} = 1.05)*

Some Magnoliales and Laurales align organ identity with spiral phyllotaxis based on "series" (Staedler et al. 2007) or pseudo-whorls (Walch and Blaise 2023), that is, sets of organs of the same type that look like whorls but are less regular because they are initiated in a spiral: the divergence angles are golden angles rather than rational angles that divide a circle into equal parts.

According to Erbar and Leins (1994), the perianth of *Liriodendron tulipifera* (Magnoliales) (K1–K9 in Figure 2.20(A)) shows a transition from spiral phyllotaxis to trimerous whorls. Three short plastochrons are followed by a relatively longer plastochron, which is followed by three short plastochrons, a longer one and again three short plastochrons. Although the perianth shows clear indications of spiral initiation, perianth organs are arranged in trimerous "whorls". The divergence angles of subsequent primordia (e.g. stamens numbered 1–8 on Figure 2.20(A)) are equal to the golden angle.

Daphnandra repandula (Laurales) has 16 tepals, eight stamens, seven staminodes and 11 carpels. The first two tepals are opposite. The other floral organs show an average divergence angle of 137.6° (±8.5°), suggesting that they are arranged in a Fibonacci spiral pattern. The perianth and androecium form pseudo-whorls of eight organs (Staedler and Endress 2009; Walch and Blaise 2023).

For both species, convergence toward the attractor (i.e. the golden angle) is slower than for Fibonacci spirals (Figure 2.15(C)) and is cyclical (Figure 2.20(B) and (D)).

2.5.7. *Lucas phyllotaxis*

The first few numbers in the Lucas sequence are 2, 1, 3, 4, 7, 11, etc. Like in the Fibonacci sequence, each number is the sum of the two previous numbers, but the initialization is different. The evolution of Lucas phyllotaxis with the decrease in plastochron ratios over time is as follows: (1, 3) → (3, 4) → (4, 7) → (7, 11) (Figure 2.13). The divergence angle between successive primordia in meristems where the numbers of parastichies are terms of the Lucas sequence is close to the Lucas angle, which is equal to:

$$\frac{5-\sqrt{5}}{10} x \, 360° \approx 99.501° \quad [2.10]$$

Marzec and Kappraff (1983) mathematically studied the effect of divergence angles on leaf distribution. They showed that certain remarkable angles optimize this distribution, meaning that they produce the densest meristems: they can all be expressed as a function of the golden ratio:

$$\tau = \frac{1+\sqrt{5}}{2} \qquad [2.11]$$

These angles are therefore all irrational, and we find among them the golden angle (137.507°) as well as the limit of the Lucas angle (99.501°):

$$137.5°{\sim}360x\frac{\tau}{1+2\tau};\ 99.5°{\sim}\ 360x\frac{\tau}{1+3\tau} \qquad [2.12]$$

Strauss et al. (2020) showed that the golden angle (137.5°) optimizes light capture but other angles, such as the Lucas angle (99.5°) do it as well, which would go against the idea that the golden angle is favored by natural selection. They noted the similarity between the effects of the divergence angle on light capture and packing efficiency in a regular lattice, as calculated by Marzec and Kappraff (1983) (see Walch and Blaise 2023, Chapter 3).

Church (1904, p. 196) lists *Sedum* (Crassulaceae), *Euphorbia* (Euphorbiaceae) and *Cereus* (Cactaceae) foliage shoot as examples of the (3, 4) mode (Church 1904, Figures 76 and 77, p. 198) and *Araucaria* (Araucariaceae) as an example of mode (7, 11), mode (4, 7) being very rare.

Rutishauser (1998) drew the cross-section of vegetative shoot apices of *Sedum sexangulare*, which has the particularity of having 18 primordia in a (3, 4) phyllotaxis. The divergence angle is circa 101° and the plastochron ratio is 1.12 (both are average values for the first 18 primordia). *S. sexangulare* got its name because most members of this species have six pseudo-orthostichies; however, some have not six but seven pseudo-orthostichies (Deschatres 1954). This is the case for the specimen examined by Rutishauser (Figure 2.21).

Since Lucas phyllotaxis is as effective as Fibonacci phyllotaxis, we wonder why the latter occurs much more frequently in nature. According to Rutishauser (2019), this predominance is not due to natural selection, but to developmental constraints ("laws of growth") that force most spiral patterns to approach the famous Fibonacci angle. Indeed, the dynamics of phyllotaxic systems must be considered.

The transition from cotyledons to Lucas phyllotaxis, or any phyllotaxis mode other than Fibonacci phyllotaxis, requires the crossing of a potential barrier to reach another potential "valley" (local minimum: see Figure 2.13). Very specific initial conditions are necessary: for example, to model the leaf apical meristem of *S. sexangulare*, the divergence angle between each of the first three organs was set at

100° (Figure 2.21). From their 1996 model, Douady and Couder also obtained (3, 4) and (4, 5) modes for specific values of the parameters.

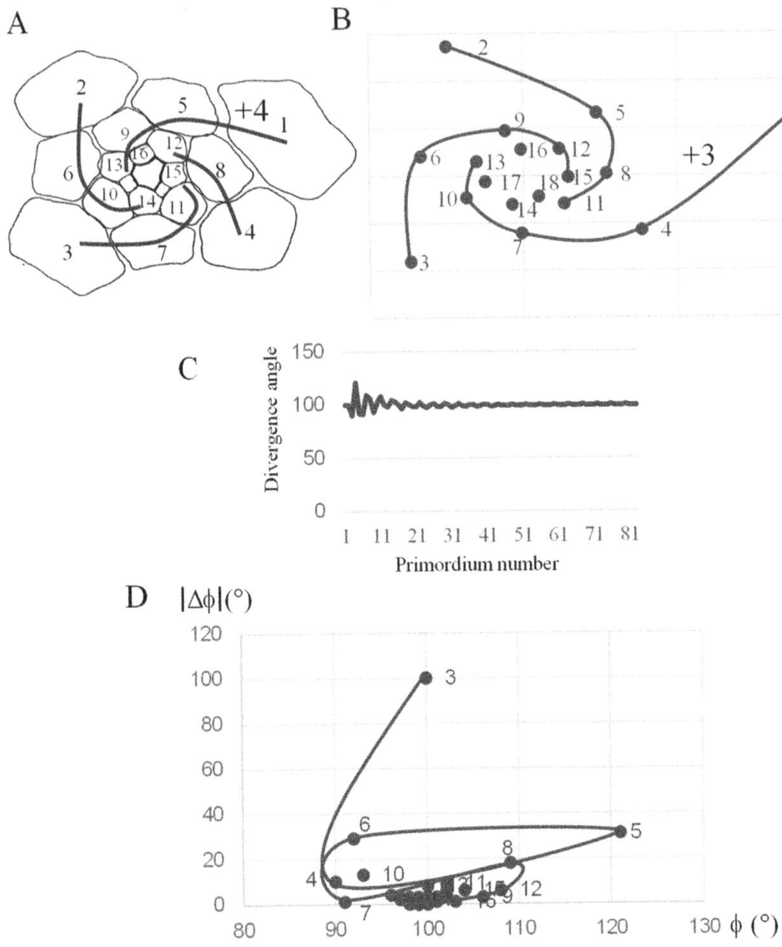

Figure 2.21. *(A) The leaf apical meristem of the specimen of Sedum sexangulare examined by Rutishauser (1998). The phyllotaxis mode is (3, 4). Only the four clockwise parastichies are shown. (B) Centric representation of organ arrangement. (C) Divergence angle converging toward the Lucas angle. (D) Phase portrait of Lucas phyllotaxis. Parameter values are* $r_0 = 0.4$, $\rho = 0.18$ *and* $r_{asympt} = 1.05$

Deschatres (1954) studied in detail the transition from a Fibonacci phyllotaxis ((2, 3) mode) to a Lucas phyllotaxis ((3, 4) mode) in *Sedum maximum* and used this as the basis for his criticism of the different "classical theories" based on divergence angles (these seems to change direction): Hofmeister's rule (1868) or the first available space (Snow and Snow 1952). From Deschatres point of view, only Plantefol's theory of multiple leaf helices (1949) would account for *Sedum maximum* specificity.

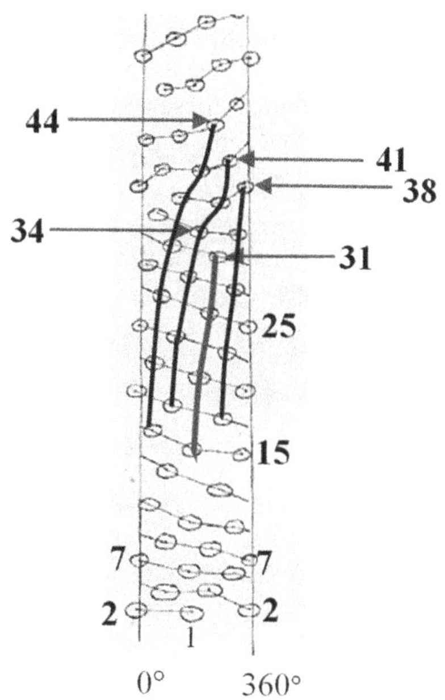

Figure 2.22. *Drawing of the stem of Sedum maximum by Deschatres (1954), with primordia numbered according to their initiation sequence. Red pseudo-orthostichy is a dead end*

Our static model (Walch and Blaise 2023, Chapter 3) allowed us to accurately calculate the average divergence angle between each leaf: around 138° at the base of the stem and 107° at the top. Pseudo-orthostichies of mode "2/5" (five leaves follow each other on a pseudo-orthostichy in two turns) were observed between leaves 6 and 26 and of mode "2/7" (seven leaves in two turns; $\phi \approx 2/7 \times 360° \approx 103°$) between leaves 38 and 52. Deschatres affirmed that: "the essential fact is the progressive transition to a disposition of the 2/7 series accompanied by an inversion of the generative spiral". According to Plantefol's theory, the "leaf-generating centers" are arranged in helices "working in synchronism" in opposite or whorled leaves and "with an offset" in "alternate" (spiral) leaves. The transition from a 2/5 series to a 2/7 series would highlight three progressively shifted leaf helices.

Careful observation of Deschatres' drawings shows that one of the pseudo-orthostichies is interrupted at leaf 31 (Figure 2.22). The transition from mode (2, 3) to mode (3, 4) is therefore abrupt and not gradual. Such transitions have also been observed by Wiss and Zagórska-Marek (2012) in *Magnolia*: a change in phyllotaxis pattern can occur with the disappearance or appearance of a parastichy, such as (8, 11) → (7, 11) or (8, 12) → (8, 13).

Zagórska-Marek (1987) attributes these transitions to discontinuities in the diameter of the apical meristem. The same type of dislocation is found in crystalline structures (edge dislocation) (Figure 2.23).

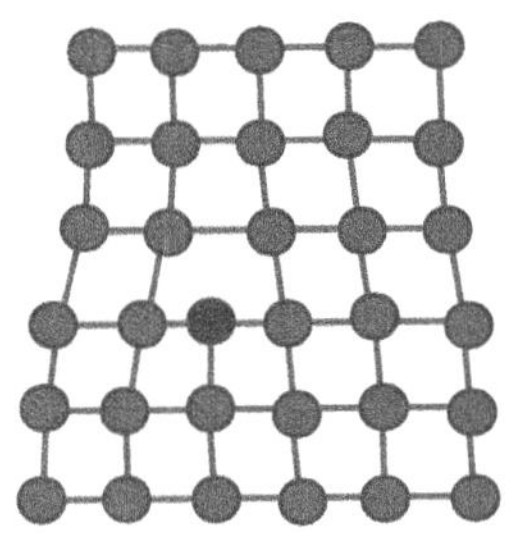

Figure 2.23. *Edge dislocation in a crystalline material*

Here, the transformation is the consequence of a discontinuity in the sequence of plastochron ratios. Where dislocation takes place, the diameter of the meristem has not had time to adapt to the decrease in the number of organs on the front, so the d/D ratio (diameter of primordia over diameter of meristem) increases suddenly.

2.6. Appendix

2.6.1. *Appendix 1: the growth index*

Let R_0 be the radius of the central zone, V the centrifugal radial velocity and T the delay between the appearance of two primordia. VT is then the distance covered by a primordium during one plastochron. The growth index $G = VT/R_0$ is a dimensionless number that characterizes the growth of the meristem irrespective of size and time scale.

If growth is exponential, then the distance from the center of a primordium is:

$$R = R_0 \exp(Vt/R) = R_0 \exp(t/TG)$$

hence:

$$r = \frac{R(t+T)}{R(t)} = \frac{e^{V(t+T)/R}}{e^{Vt/R}} = \exp(VT/R) = \exp(G)$$

and the growth index G = ln(r).

2.6.2. *Appendix 2: cylindrical versus centric lattices*

The point K of coordinates (θk, hk) on a cylindrical lattice can be expressed by the complex number:

$$z_k = (-\phi + iG)k$$

where k is the node number.

Applying the conformal transformation:

$$Z_k = e^{-iz_k} = e^{Gk}e^{ik\varphi}$$

we obtain the algebraic form of the complex number Z_k:

$$Re(Z_k) = e^{Gk}\cos(2\pi k\varphi)$$

$$Im(Z_k) = e^{Gk}\sin(2\pi k\varphi)$$

Primordia are located on a logarithmic spiral with the parametric equation:

$$R_k = r^k \approx e^{kG}$$

$$\vartheta_k = 2\pi k\varphi$$

We eliminate k to obtain the polar equation of the spiral:

$$R = \exp(\frac{G}{\varphi}\vartheta)$$

This logarithmic spiral is invariant for any transformation (similarity) consisting of a rotation of center O and angle a and a scaling of factor:

$$\rho = \exp(Ga/\phi).$$

By this conformal transformation, a translation within the cylindrical lattice:

$$z'_k = (\varphi + \varphi_{a)}k + i(G + G_a)k$$

is transformed into a similarity within the centric lattice:

$$Z'_k = e^{(G+G_a)k} e^{i(\varphi+\varphi_a)k}$$

3

Symmetry in Ranunculales

The order is highly diverse in terms of floral structure. As such, it occupies a transitional position between basal angiosperms and core eudicots. Ranunculales is the sister clade to all other eudicots. Among the seven families within this order, independent transitions toward zygomorphy have occurred in Papaveracaeae and Ranunculaceae (coincidently with the evolution of spurs) and in Menispermaceae. Spurs develop as outgrowths of perianth organs (either petals or sepals or both) relatively late during flower development, with a function of storing nectar. The size and shape of spurs may play a role in selecting pollinators according to size and morphology criteria. Spurs evolved multiple times in angiosperms and are considered floral key innovations because their presence, at least in some groups, is associated with species diversification (Damerval and Nadot 2007; Hodges and Arnold 1995).

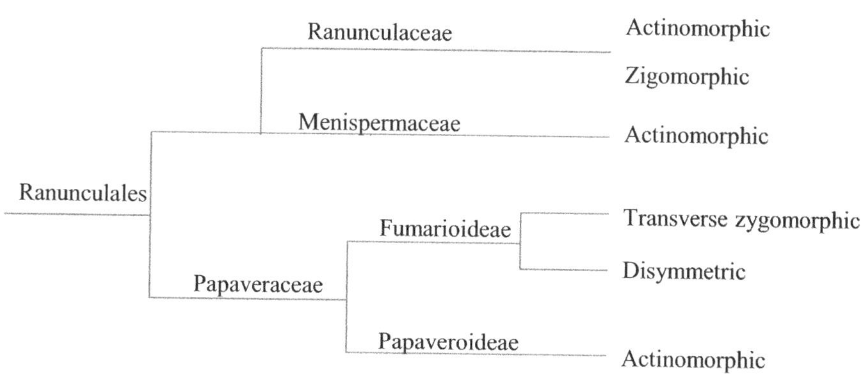

Figure 3.1. *Partial phylogenetic tree of studied Ranunculales*

For a color version of all the figures in this chapter, see www.iste.co.uk/walch/phyllotaxis2.zip.

3.1. Papaveraceae

In Papaveraceae, *Chelidonium majus* and *Lamprocapnos spectabilis* both have two axes of symmetry: the dorsoventral and the transverse one. However, with respect to these axes, *C. majus* is radially symmetric (its shape is "circular", Figure 3.2(B)) while *L. spectabilis* is not (its shape is "elliptical"; Figure 3.5(B)). *C. majus* is actinomorphic, while *L. spectabilis* is disymmetric. Actinomorphy is the ancestral state in the early diverging eudicot family Papaveraceae. Zygomorphy evolved once in the subfamily Fumarioideae (e.g. *Capnoides sempervirens*, which has lost one symmetry axis; Figure 3.6) from a disymmetric state. Unusual within angiosperms, this zygomorphy takes place along the transverse axis of the flower, which is asymmetry with respect to the dorsoventral axis (Damerval et al. 2013).

Actinomorphic Papaveraceae mature flowers have neither nectary nor spur, while species with disymmetric or zygomorphic mature flowers have spurs and nectaries borne at the abaxial (outer) base of stamen filaments.

Expression analysis of *CYC*-like genes was undertaken by Damerval et al. (2007, 2013) in three representative species of each of the three types of floral symmetry encountered in the Papaveraceae family: *Chelidonium majus* for actinomorphic, *Lamprocapnos spectabilis* for disymmetric and *Capnoides sempervirens* for zygomorphic flowers.

3.1.1. *Papaveroideae*

Chelidonium majus

The inflorescence of *Chelidonium majus* (Papaveroideae) is composed of umbelliform cymes. Flowers are formed in spirals subtended by a bract (Figure 3.2(A)). The initiation of the two sepals is simultaneous, compared to that of two pairs of petals. The first whorl of stamens is initiated in pairs, then a second whorl of four stamens is initiated. All cycles alternate one after the other. The actinomorphy of the flower is established from the initiation of the petals and persists throughout development until maturity (Payer 1857; Figure 3.2(B)).

For modeling a lateral flower, the source of adaxial inhibition is the terminal flower, the abaxial source being the bract (Figure 3.2(A); Table 3.1).

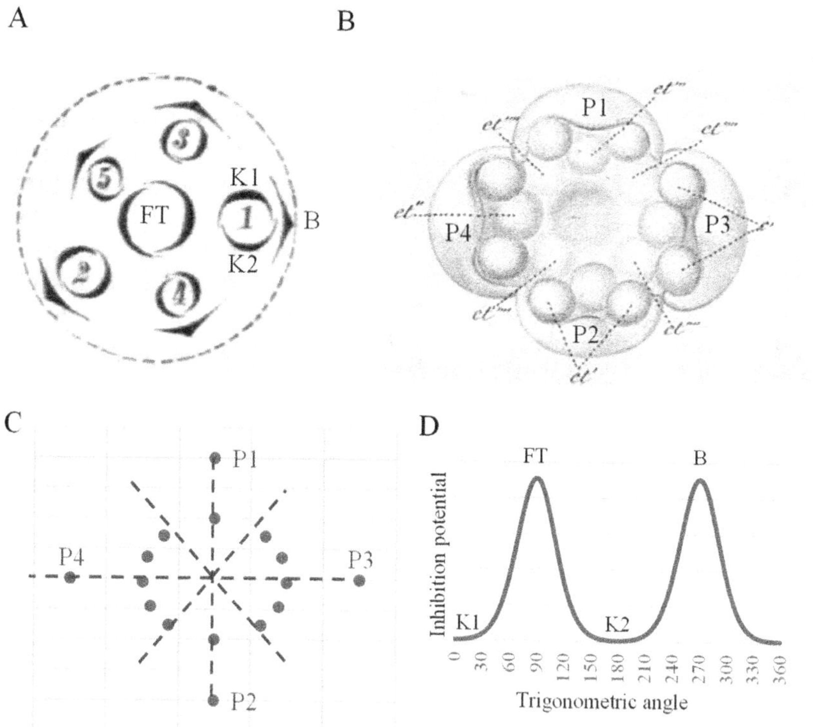

Figure 3.2. *Chelidonium majus. (A) Inflorescence diagram from Eichler (1878); FT = terminal flower; B = bract; numbers = lateral flowers. (B) Corolla and androecium according to Payer (1857). (C) Model of lateral flower. Unlabeled points are stamens. (D) Inhibition profile at K1 initiation of lateral flower*

Organ	Angle (°)	Plastochron ratio
Terminal flower	0	1
Bract	270	3
K1		1
K2		3
P1		1
P2		1
P3		1
P4		2
S		1
S		...

Table 3.1. *Chelidonium majus modeling parameters*

3.1.2. *Fumarioideae*

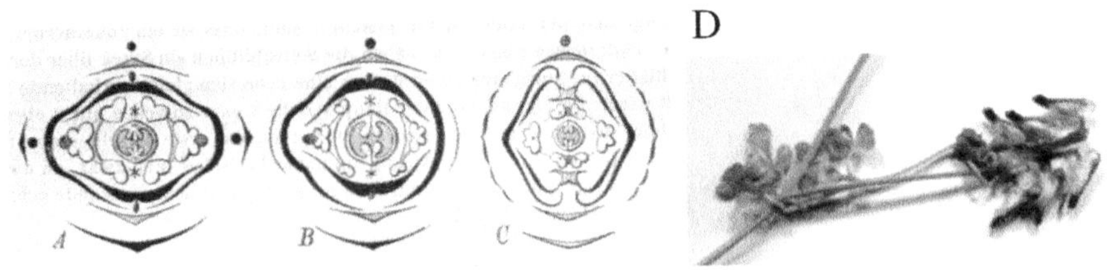

Figure 3.3. *Floral diagrams according to Eichler (1878). (A) Lamprocapnos formosa (syn. Dicentra): disymmetric. (B) Capnoides cava (syn. Corydalis): transverse monosymmetric. (C) Hypecoum procumbens: disymmetric. (D) Mature flowers of zygomorphic Fumaria capreolata (Flore virtuelle d'Orsay)*

Disymmetry refers to flowers with **only** two different orthogonal symmetry axes (Endress 1999). In Fumarioideae, the two symmetry axes are, respectively, the dorsoventral and transverse axes. In the disymmetric *Lamprocapnos*, two spurs are formed from the two outer petals in the transverse axis, giving the flower a heart-shape, hence the common name "bleeding hearts" (Figure 3.3(A)). In zygomorphic (monosymmetric) species, only one spur is formed and a 90° resupination puts this single spur in a vertical orientation, mimicking a dorsoventral zygomorphy at maturity (Damerval et al. 2013) (Figure 3.3(B)).

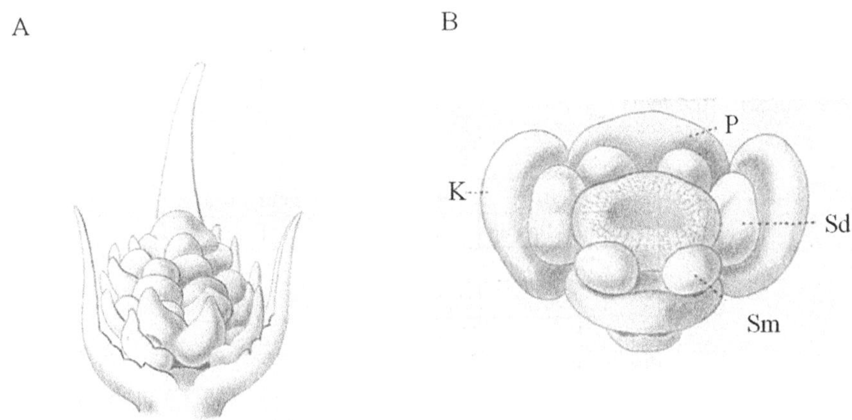

Figure 3.4. *Platycapnos spicata (Papaveraceae) according to Payer (1857). (A) Inflorescence. (B) Bud after stamens initiation showing monothecal (dorsal and ventral) and dithecal (transverse) stamens. Sm = monothecal stamen; Sd = dithecal stamen*

Lamprocapnos spectabilis

In *Lamprocapnos spectabilis* (Fumarioideae), sepals are initiated in dorsoventral position, and the two cycles of petals appear in alternate positions. Outer and inner petals are morphologically different from inception onward. The androecium consists of six stamens organized in two triplets along the transverse plane, with the central stamen of each triplet opposed to the outer petal. The central stamen primordia form two thecae (dithecal stamens); whereas, the lateral ones form only one theca (monothecal stamens; Figure 3.4(B)). A gynoecium ring primordium forms at the center of the apex. Therefore, at the end of ontogenesis, all floral cycles appear disymmetric (Damerval et al. 2013).

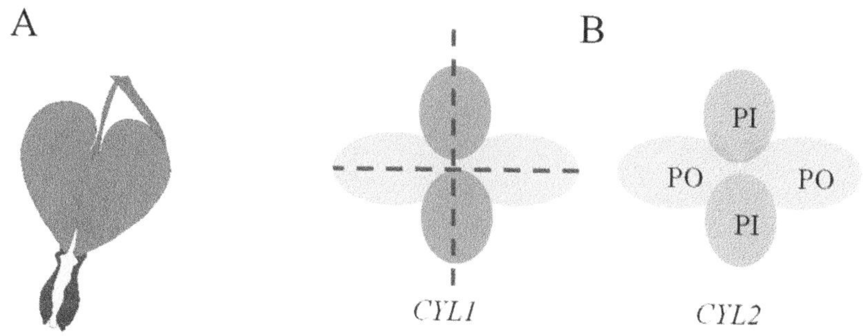

Figure 3.5. *Lamprocapnos spectabilis redrawn from Damerval et al. (2013). (A) Mature flower. (B and C) Diagrams of LsCYL1 and LsCYL2 genes expression in petals from Damerval et al. (2013). PO = outer petal; PI = inner petal; Dark blue = strong expression; light blue = weak expression. The shape of the flower is stylized to represent the symmetry of the corolla at anthesis. Dashed lines = axes of symmetry*

Damerval et al. (2007) have characterized two *CYC*-like paralogous lineages in Papaveraceae (*PapCYL* 1 and *PapCYL2*), resulting from a duplication predating the divergence between Fumarioideae and Papaveroideae. Damerval et al. (2013) investigated *CYC*-like genes expression from *L. spectabilis* by reverse transcriptase-polymerase chain reaction (RT-PCR). Expression of *LsCYL1* and *LsCYL2* has strong signals in sepals and inner petals and a weak signal in outer petals (Figure 3.5(B)). The disymmetry of the flower at anthesis is in accordance with that of the expression of *CYC*-like genes (Curie's principle).

Capnoides sempervirens

In *Capnoides sempervirens*, transverse asymmetric growth starts with outgrowth of one of the two dithecal stamen filaments (see Figure 3.4(B)) that will bear the only nectary. This asymmetry culminates with the formation of an outer petal (P2) spur enclosing this latter stamen filament (Figures 3.3(B) and 3.6(B)). This spur is formed transversally and a 90° resupination puts this single spur in a vertical orientation, mimicking a dorsoventral zygomorphy at maturity (Damerval et al. 2013).

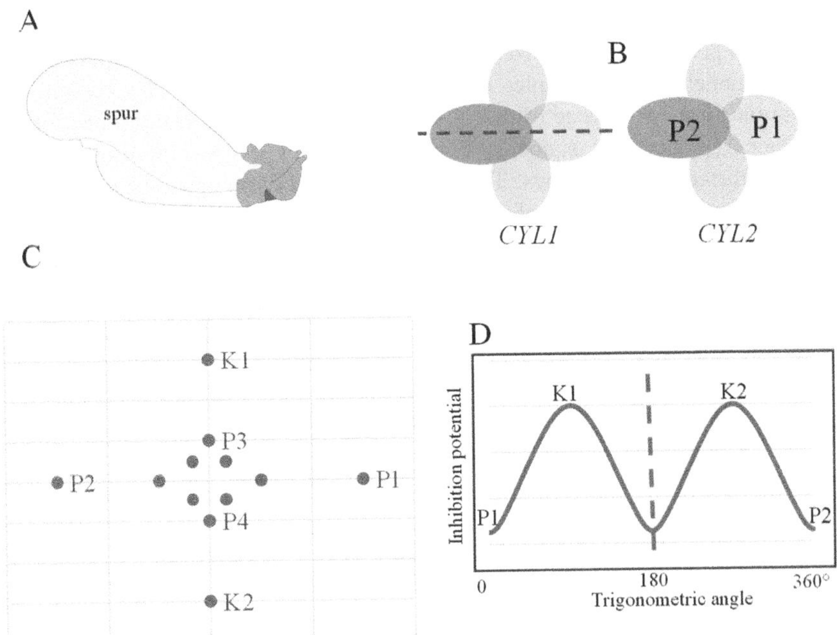

Figure 3.6. *Capnoides sempervirens redrawn from Damerval et al. (2013). (A) Mature flower. (B) Expression of CsCYL1 and CsCYL2 genes in petals according to Damerval et al. (2013). Dark blue = strong expression; light blue = weak expression. (C) Model of Capnoides flower. Unlabeled organs = stamens. (D) Profile of the inhibition potential at the initiation of external petals (Landau potential). Dashed line = symmetry axis of the figure*

In *Capnoides sempervirens*, the expression of both *CYC* homologs (*Pap-CYL*) appears to be weaker in the basal region of the outer petal that does not develop a spur (P1), than in the corresponding region of the outer petal that develops a spur (P2) (Figure 3.6(B)). *CYC-like* expression is also detected in both dithecal stamen filaments, but more strongly in the single outgrowth that will form the nectary. The shift from disymmetry to zygomorphy in Fumarioideae is associated with the asymmetric growth of stamen filaments, followed by the asymmetric development of the nectar outgrowth and the spur along the transverse plane. These results show that members of the *CYC* family have an expression pattern asymmetric with respect to the dorsoventral axis at late stages of development in Fumarioideae, when zygomorphy is already discernible in the bud (Damerval et al. 2013).

> It is worth underlining that at the scale of the inflorescence in *C. sempervirens*, there seems to be no rule **for which** the transverse sides of the flower will form the nectary outgrowth and spur. This suggests that early random fluctuations in some morphogenetic cue along the transverse plane might initiate asymmetric gene expression and growth, and that the process, once engaged, becomes fixed (Damerval et al. 2013, p. 399).

Damerval et al. (2013) describe, without knowing, a second-order phase transition.

Organ	Angle (°)	Plastochron ratio
K1	90	1
K2	270	2
P1		1
P2		1.5
P3		1
P4		2
S1		1
S		…

Table 3.2. *Fumarioideae modeling parameters*

This type of bifurcation of a potential is well known in physics under the name of Hopf bifurcation, which applies to second-order phase transitions (different from the first order, such as melting of ice or boiling of water) with a change in the symmetry of the system, for example, the Rayleigh–Bénard convection. If you heat a liquid from below (like in a saucepan), at the beginning, the temperature difference between the hot plate and the interface between the liquid and the ambient air is small: all the heat produced by the gas cooker heats the liquid layer without macroscopic movement, by only invisible agitation of the liquid molecules (conduction phenomenon). Beyond a certain temperature difference between the hot plate and the ambient air, macroscopic convective cells are formed in the liquid which transport heat more quickly from the hot source to the cold source (convection phenomenon; Figure 3.7).

This phenomenon was discovered by Henri Bénard (*Les Tourbillons cellulaires dans une nappe liquide propageant de la chaleur par convection en régime permanent* 1900). The temperature difference between hot source and cold source necessary for the macroscopic movement of the fluid layer was calculated by Rayleigh (1916). The convective cells can rotate clockwise or counterclockwise (Figure 3.7), and these two directions of rotation have equal probabilities: the symmetry of the experiment dictates that neither direction can be privileged over the other.

Figure 3.7. *Convective cells (Bénard)*

At the initiation of the external petals (P1, P2), the inhibition potential generated by the K1 and K2 sepals is symmetrical with respect to the dorsoventral axis. The two minimum positions, at 0° and 180°, are equiprobable because their level is equal due to symmetry (Figure 3.6(D)). In the case of *Lamprocapnos*, the two outer petals are initiated simultaneously, and the symmetry is preserved. In the case of *Capnoides*, one of the two positions is randomly selected. The probabilities of the two positions are equal due to the equality of the potential minima. This is what Damerval et al. (2013) described.

Second-order phase transitions "break" the symmetry: the phase with transition (zygomorphic *Capnoides*: one axis of symmetry) is less symmetrical than the phase without transition (disymmetric *Lamprocapnos*: two axes of symmetry).

3.2. Menispermaceae

Payer (1857) followed all the phases of the organogenesis of *Menispermum canadense* (Menispermaceae) flower "which bloom every year in great abundance at the Museum (Paris)".

> The inflorescence of *Menispermum* is a biparous cyme. In the axil of each secondary lateral bract (bracteole), six sepals are initiated in two whorls of three each. They rise successively on the floral mamelons, one of the anterior first, then the posterior and finally the other anterior. In the bud, they remain small and never overlap. The three sepals of the inner whorl alternate with the three sepals of the outer whorl, which is to say there is one on the side of the 'mother bract' and two on the opposite side.
>
> The corolla of *Menispermum* is also composed of two whorls of three petals each. These two whorls alternate with each other and with the two whorls of the calyx. These petals, once initiated, grow only very slowly at the beginning, so that the stamens, although initiated later, quickly exceed them by much, and if one did not follow step by step

the development of these petals and stamens, one would think that the petals only appear after the stamens and are even dependent on them (Payer 1857). *Menispermum* is actinomorphic.

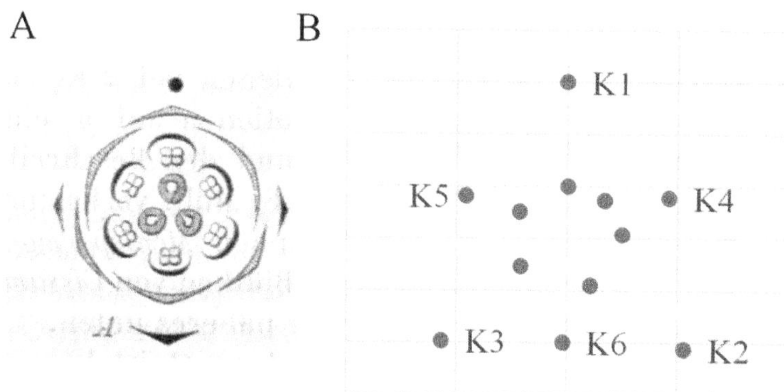

Figure 3.8. *(A) Cocculus carolinus (Menispermaceae) floral diagram (Eichler 1878). (B) Cocculus model. Unlabeled primordia = petals*

Organ	Angle (°)	Distance to the center
Adaxial source	90	3
B	270	2
α	0	2
β	180	2
		Plastochron ratio
K1		1
K2		1
K3		1.5
K4		1
K5		1
K6		2
P		1
P		1
P		1
P		1
P		1
P		1

Table 3.3. *Cocculus modeling parameters*

For modeling, there is no need to "force" symmetry (see Chapter 4). The plastochron ratio of K3 corresponds to the transition from the first sepal whorl of Menispermaceae to the second whorl (Table 3.3).

3.3. Ranunculaceae

Ranunculaceae, with more than 2,500 species, is the largest family in the Ranunculales order. Inflorescences are variable and consist of terminal single flowers (*Ranunculus ficaria*), dichasia (*Clematis*) or racemes (*Actea* and *Aconitum*). This family is especially remarkable by the diversity of its perianth form and architecture, exhibiting variable merism (2- to 5-merism), organ number (determinate or variable), phyllotaxis (spiral or whorled) and perianth differentiation (unipartite or bipartite with inner organs [petals] morphologically differentiated from outer ones [sepals]). The single evolutionary transition from actinomorphy (or pseudo-actinomorphy, in the case of spirally inserted organs) to zygomorphy took place in the stem lineage of the species-rich Delphinieae tribe (accounting for 26% of all Ranunculaceae species) and has involved the asymmetric formation of spurs in both sepals and petals (Jabbour et al. 2014; Ronse de Craene 2022).

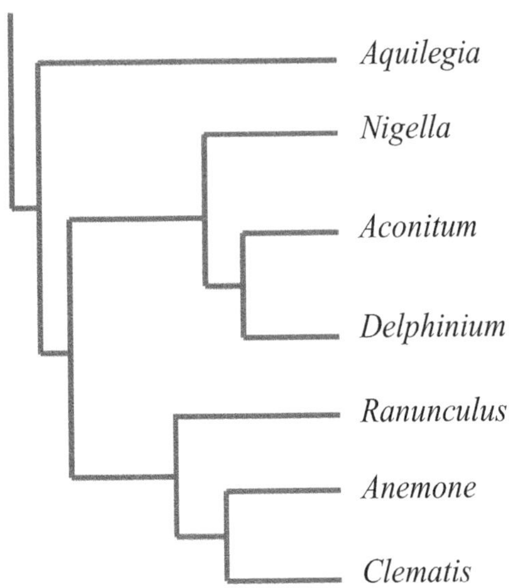

Figure 3.9. *Phylogenetic tree of the species studied (from Carrive et al. (2020))*

Jabbour et al. (2014) characterized *CYC* homologs in 18 genera of Ranunculaceae, in a sampling that represents the floral morphological diversity of this tribe and reconstructed the evolutionary history of this gene family in Ranunculaceae. Within each of the two *RanaCYL* (Ranunculaceae CYCLOIDEA-like) lineages previously identified, an additional duplication possibly predating the emergence of the Delphinieae was found, resulting in up to four gene copies in zygomorphic species. Expression analyses indicate that the *RanaCYL* paralogs are expressed early in floral buds and that the duration of their expression varies between species and paralog class.

Aquilegia

The development of the lateral flowers of *Aquilegia vulgaris* is **basipetal**. The terminal flower develops first, and then the axillary buds located just below the terminal flower develop before those located lower down (Claßen-Bockhoff and Bull-Hereñu 2013).

The floral meristem of *Aquilegia* successively initiates two bracteoles α and β (Figure 3.10(C)), and then five sepals in spiral symmetry (quincuncial; Figure 3.10(D)), which becomes valvate at anthesis by contact pressure (Figure 3.10(B); see Figure 3.15). All five petal primordia are initiated simultaneously as a pentamerous whorl of organs in positions alternating with sepals. Then the stamens form 10 orthostichies, starting from the five sepals and the five petals (Figure 3.10(E)). The orthostichies have five to six stamens each, all of equal size. Then five carpel primordia are initiated (Tucker and Hodges 2005).

For the modeling, the bracteoles α and β were arranged according to the Fibonacci angle (around 140°), the two bracteoles being at very unequal distances from the center of the flower, which generates a quincuncial arrangement of the sepals. Then a high plastochron ratio ($r = 1.3$) before the initiation of the corolla allows it to form a pentamerous whorl alternating with that of the calyx. Also, the perfect orthostichies of the androecium are generated by unitary plastochron ratios within each whorl and high plastochron ratio ($r = 1.5$) between two whorls; otherwise, the stamens would form spirals.

The mature *Aquilegia* flower is actinomorphic, including its nectariferous petal spurs. Once all organs are initiated, the corolla forms five spurs that extend into the third dimension (Figure 3.10(A)). Whittall and Hodges (2007) showed that the depth of these spurs increased with the length of the tongue of the pollinators (around 20 mm for these *bombus*).

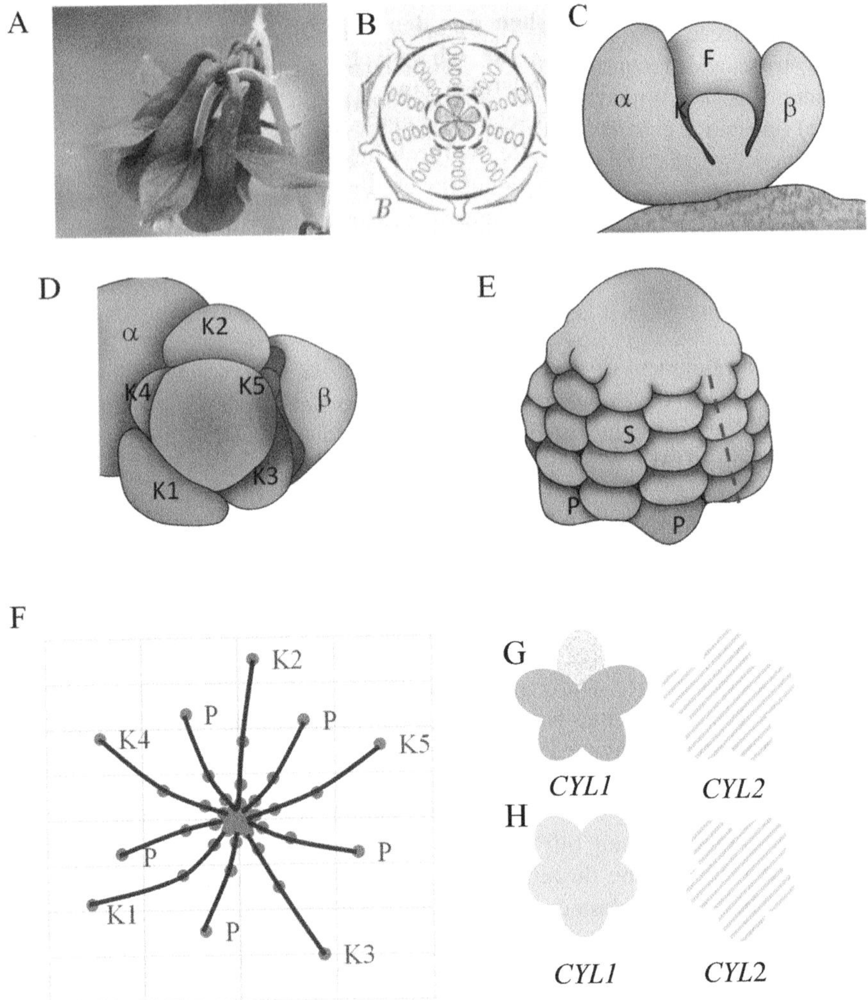

Figure 3.10. *(A) Aquilegia vulgaris: mature flower showing five spurs (photo from Florian Jabbour) and (B) floral diagram (Eichler 1878). (C and D) Initiation of bracteoles α and β and sepals of A. olympica, according to Tucker and Hodges (2005). (E) Floral meristem showing the orthostichies (dashed red line) composed of stamens according to Tucker and Hodges (2005). (F) Model of Aquilegia vulgaris showing 10 stamen orthostichies. Unlabeled points = stamens. Expression of CYL-like genes in the developing meristem of A. alpina according to Jabbour et al. (2014): (G) Calix and (H) Corolla*

Organ	Angle (°)	Distance to the center
B	270	3
α	140	2.7
β	0	1.7
		Plastochron ratio
K1		1.02
K2		1.02
K3		1.02
K4		1.02
K5		1.3
P1		1
P2		1
P3		1
P4		1
P5		1.5
S		1
		...

Table 3.4. *Aquilegia vulgaris modeling parameters*

The expression of *CYL* genes in *Aquilegia* is radially symmetric (ubiquitous or lacking), except for *AaCYL1* on the calyx, which is not actinomorphic but spirally symmetric (Jabbour et al. 2014) (Figures 3.10(G–H)).

Caltha

Caltha palustris (Ranunculaceae), known as marsh-marigold or kingcup, is native to marshes, fens, ditches and wet woodlands in temperate regions of the Northern Hemisphere. Eichler (1878) drew a diagram of a partial inflorescence of a lateral branch of *C. palustris* with a terminal flower (F1) and two lateral flowers (F2, F2'; Figure 3.11(B)). The terminal flower has two bracteoles, α and β, in the axils of which the lateral flowers F2 and F2' arise. Flower F2 (in the axil of α) has one bracteole, while F2' (in the axil of β) has none. The K4 sepal of the F2' flower is adaxial.

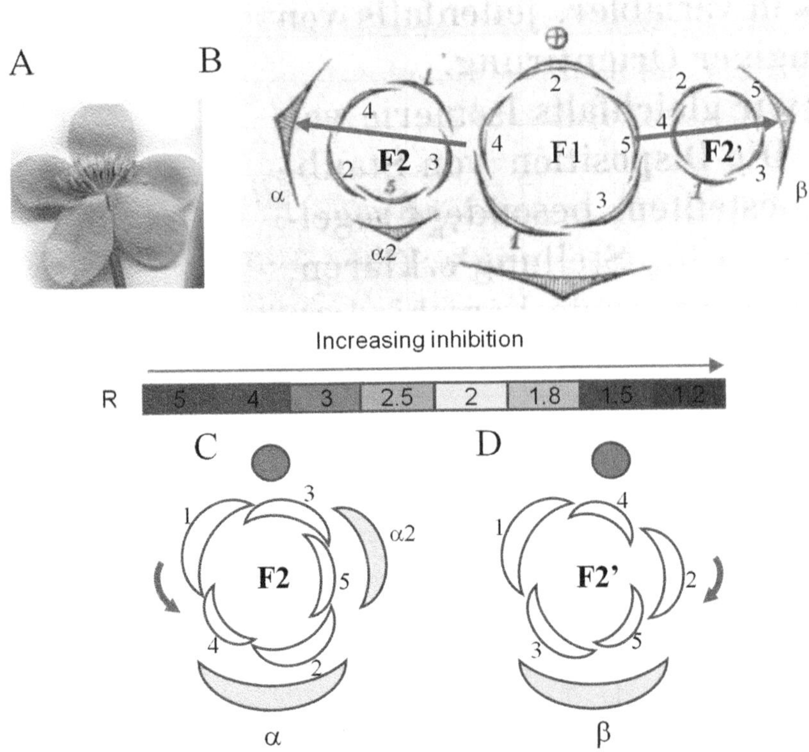

Figure 3.11. *Caltha palustris. (A) Mature flower from Flore virtuelle d'Orsay. (B) Diagram of a partial inflorescence (Eichler 1878). Model of sepal initiation in C. palustris (C) in the F2 flower and (D) in the F2' flower, whose arrangement is of the Primulaceae type (K4 is adaxial). Blue circle = adaxial repulsion source. Arrows: the two flowers are antidromous. The color represents the apparent distances from the sources of inhibition to the apex (R), the intensity of the inhibition being all the more important as the distance from the organ to the apex is low (see Figure 1.4)*

Sepals of the terminal flower have a main quincuncial arrangement, which is determined by the two bracteoles α and β. Sepals in the lateral flowers of *C. palustris* cannot be in a "main quincuncial arrangement" (K2 adaxial) because they do not have two bracteoles. The first sepal of F2 and F2' is located on the dorsal side (and not on the ventral side, as in the main quincuncial arrangement) away from the bract and the single bracteole for F2. Modeling of these arrangements requires an adaxial source of "repulsion" for F2 and one that is slightly offset for F2'. These sources can be located near the point of contact between the lateral flower and the terminal flower (adaxial circle on Figure 3.11(C) and (D)).

The terminal flower develops before the lateral flowers. Therefore, the α and β bracteoles must already have played their role as a source of inhibition during sepal initiation in F1, before being pushed back by the sympodial meristems of the lateral flowers and becoming their bracts. We can think that the adaxial and abaxial repulsions influencing sepal initiation in F2 and F2' are not caused by "auxin" concentration, but rather by contact pressure exerted, on the one hand, by the flower F1 and, on the other hand, by the bract α or β.

Nigella

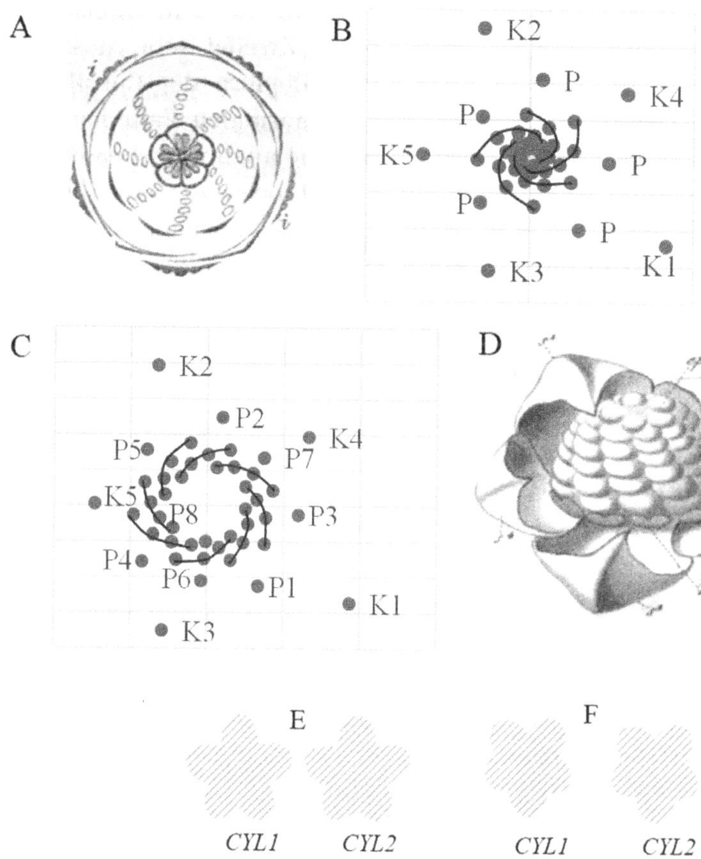

Figure 3.12. *Nigella damascena. (A) Floral diagram of octomerous flower (Eichler 1878). i = involucrum. (B) Model of pentamerous flower. (C) Model of octomerous flower. (D) Nigella sativa according to Payer (1857). Lack of expression of CYL-like genes in the developing meristem of pentamerous Nigella damascena, according to Jabbour et al. (2014). (E) Calix. (F) Corolla*

Nigella perianth is actinomorphic at anthesis. The perianth is composed of five petaloid sepals and five or eight petals, then stamen being arranged in spirals (Figure 3.12).

	Pentamerous flower		Octomerous flower	
Angle (°)	**Organ**	**Distance to the center**	**Organ**	**Distance to the center**
0	I1	1	I1	1
72	I2	1	I2	1
144	I3	1	I3	1
216	I4	1	I4	1
288	I5	1.12	I5	1.12
		Plastochron ratio		**Plastochron ratio**
	K1	1.12	K1	1.12
	K2	1.12	K2	1.12
	K3	1.12	K3	1.12
	K4	1.12	K4	1.12
	K5	1.12	K5	1.12
	P	1.1	P	1.045
	P	1.1	P	1.045
	P	1.1	P	1.045
	P	1.1	P	1.045
	P	1.1	P	1.045
	S	1.06	P	1.045
	…	…	P	1.045
			P	1.045
			S	1.02
			…	…

Table 3.5. *Nigella damascena modeling parameters. Left table: pentamerous flower; right table: octomerous flower*

Compared to *Aquilegia* (Table 3.4), a lower plastochron ratio between the androecium and the corolla, as well as plastochron ratios greater than 1 for spiral symmetry between stamens ensure the propagation of the spiral initiation from the calyx to the last stamens. The difference between the octomerous flower and the pentamerous flower is only smaller plastochron ratios for the petals and stamens (Table 3.5).

Both *NdCYL1* and *NdCYL2* genes were found to be expressed in floral buds. Expression was also detected in leaves. Their expression level in flowers decreases over time. In 3–5 mm diameter buds, *NdCYL1* is faintly expressed in the dorsal sepal, and *NdCYL2* expression is restricted to the stamens; no expression was detected in any organ of 6–7 mm diameter buds (Jabbour et al. 2014).

Delphinium

The Ranunculaceae family includes four highly zygomorphic genera: *Aconitum*, *Delphinium*, *Gymnaconitum* and *Staphisagria*. Zygomorphy in Delphinieae is late, occurring after the end of organogenesis (Jabbour et al. 2009) (Figure 5.1).

The raceme of *Delphinium*, like that of *Aconitum*, is remarkable for the presence of sterile bracteoles. The inflorescence meristem produces (in this order) bracts and floral meristems that initiate two lateral bracteoles (Figure 3.13(C)). These latter become different in size shortly after inception (α and β). Five sepal primordia successively initiate in a spiral manner clockwise or counterclockwise (with ca. 137.5° between two consecutive primordia) and then gradually grow into crescent-shaped organs. K1 is closer to α and K2 is adaxial (typical quincuncial aestivation; Figure 3.13(D)). Petal primordia (generally eight) are spirally initiated (Figure 3.13(E)). After initiation of the last petal primordium, stamen primordia are subsequently spiral produced.

In *Delphinium* and *Aconitum*, late in development, the two dorsal petals (P2 and P5) form a spur that is placed inside the spurred adaxial sepal (K2): at anthesis, they point backwards in the third dimension. The flower becomes zygomorphic (Figures 3.13(A) and (B)).

Delphinium species with the shortest nectar spurs are pollinated by bees or bumblebees, those with intermediate nectar spurs by hummingbirds, and those with the longest nectar spurs by hawkmoths. In *Delphinium*, flat petals also act as pollinator filters and/or attractors, in addition to the spurred petals. They are used as a visual cue for pollinators, due to the hairs that turn yellow at maturity (Jabbour et al. 2009).

For the modeling (Figure 3.13(B)), the angular positions of the three inhibition sources outside the floral meristem were selected in accordance with the SEM (Figure 3.13(D); α and β are reversed compared to Figure 3.13(B); B = 270° for bract; α = 40°; β = 184° for bracteoles). The inhibition potentials produced by the two bracteoles must be different enough to produce a quincuncial calyx ($R\alpha = 2.2$ and $R\beta = 1.5$).

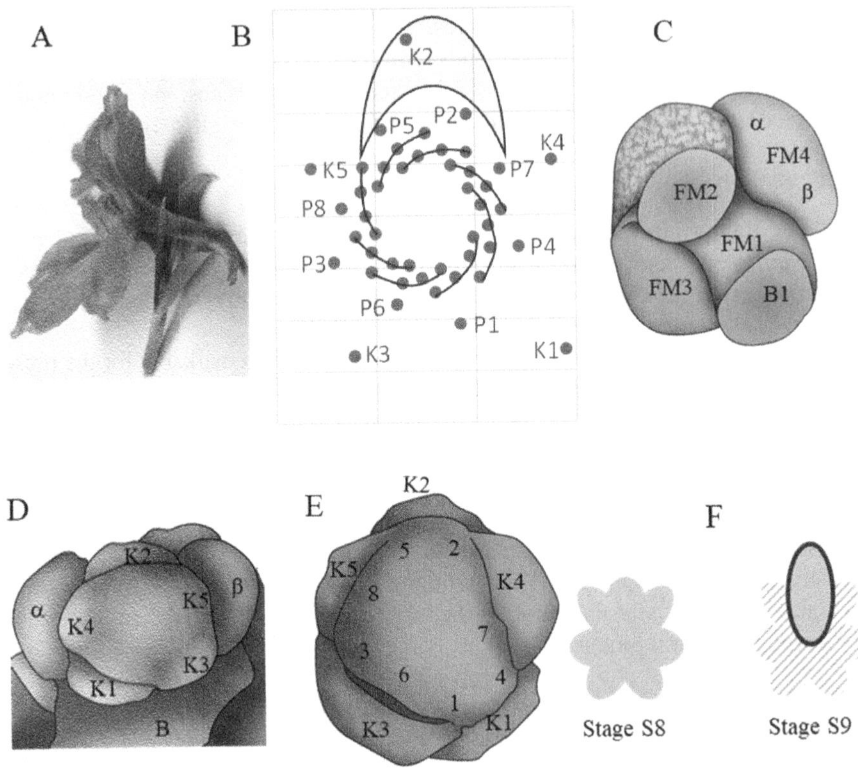

Figure 3.13. *(A) Delphinium grandiflorum showing spur (photo from Florian Jabbour). (B) Model of Delphinium grandiflorum that fits exactly the floral diagram of Jabbour et al. (2009). (C) Initiation in inflorescence meristem of bracts, floral meristems and bracteoles of Aconitum taipeicum (Chang et al. 2019). The oldest primordium (FM4) has already developed its bracteoles. (D) Initiation of sepals in Delphinium anthriscifolium (Zhang et al. 2022b). (E) Initiation of eight petals in Delphinium giraldii, according to Chang et al. (2019). (F) Expression of DeajCYC2B gene and symmetry of the corolla in mature flowers of Delphinium ajacis, according to Zhao et al. (2023). Dark blue = strong expression, light blue = weak expression*

In *Aconitum*, *Delphinium* and *Nigella*, the five sepals are initiated in two revolutions of spirals, that is, a mode (2, 3), and the petals in three revolutions, that is, a mode (3, 5), which corresponds to a decrease of the plastochron ratio of a Fibonacci spiral (Jabbour et al. 2009) (Table 3.6).

Organ	Angle (°)	Distance to the center
B	270	3
α	40	2.2
β	184	1.5
		Plastochron ratio
K1		1.116
K2		1.116
K3		1.116
K4		1.116
K5		1.116
P1		1.04
P2		1.04
P3		1.04
P4		1.04
P5		1.04
P6		1.04
P7		1.04
P8		1.04
S1		1.015

Table 3.6. *Delphinium grandiflorum modeling parameters*

Zhao et al. (2023) conducted detailed gene expression and functional studies in *Delphinium ajacis*, a species that produces flowers with one single-spurred dorsal sepal, four non-spurred sepals, one fused single-spurred dorsal petal, six reduced petals, 13–17 stamens and one carpel.

Two *CYC*-like genes, *DeajCYC2B* and *DeajCYC2A*, have evolved to define the dorsal and lateral-ventral identities of the flower, respectively. *DeajCYC2B* was initially expressed in floral meristems and all floral organ primordia and showed dorsal petal-specific expression at S9, when the sepal spur became visible (Figure 3.13(F)). Knockdown using VIGS (virus-induced gene silencing) of *DeajCYC2B* led to a reduction in spurred petal formation: while the genes controlling spur formation remain to be identified, it is already clear that the dorsal identity gene *DeajCYC2B* promotes the development of spurs (Zhao et al. 2023).

Floral symmetry is prepatterned by the influence on the perianth arrangement of organs peripheral to the floral meristem (bract, bracteoles). At the late stage of development, the position of the calyx primordia can no longer be modified, only their size changing significantly, and the expression of *CYC*-like genes late becomes asymmetric with respect to the transverse axis. The flower acquires a positional zygomorphy (Figure 3.13(B); see Chapter 5).

Ranunculus repens (Fibonacci)

The floral diagram of *Ranunculus repens* drawn by Meicenheimer (1979) shows that bracteoles and sepals are on the same Fibonacci spiral; whereas, petals are arranged in a pentamerous whorl; the spiral then continues with the stamens and carpels (Figure 3.14(B)). Meicenheimer published one of the few available plastochron data sets measured on real flowers. These plastochron ratios were used as parameter values in our model (Table 3.7).

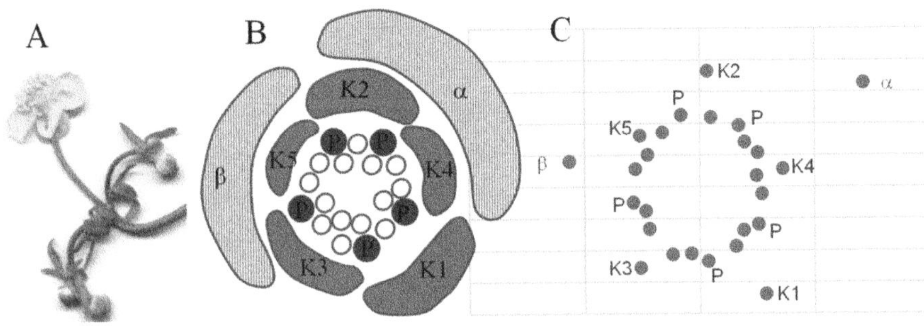

Figure 3.14. *(A) Biparous cyme of Ranunculus bulbosus from Flore virtuelle d'Orsay. The corolla is actinomorphic at anthesis. (B) Floral diagram of Ranunculus repens, redrawn from Meicenheimer (1979). (C) Model of the floral meristem of R. repens using the plastochron ratios measured by Meicenheimer as parameter values. α, β = bracteoles; K = sepals; P = petals. Unlabeled points = stamens*

Cunnell (1958, 1960) studied the aestivations (preflowering) of *Ranunculus repens*. The calyx is perfectly quincuncial. The initiation of the five petals being perfectly synchronous, the corolla could be valvate, but because the inhibition field produced by the sepals is spiral, it could also be quincuncial. The overlap of the petals depends on a random movement, according to Cunnell.

Due to the symmetry of the different aestivations, there are 10 possible combinations (five dextral and five sinistral) for quincuncial, cochleate and imbricate aestivations, and only two for contorted aestivation (one dextral and one sinistral). Indeed, in the latter case, all 1/5 turn rotations reproduce the same arrangement (Figure 3.15(C)).

Organ	Angle (°)	Distance to the center
B	270	3
b1	30	2
b2	170	1.4
		Plastochron ratio
K1		1.116
K2		1.116
K3		1.116
K4		1.116
K5		1.116
P1		1
P2		1
P3		1
P4		1
P5		1.012
S		1.012

Table 3.7. *Ranunculus repens modeling parameters from Meicenheimer (1979)*

Let us hypothesize that the transition from spiral initiation to whorled aestivation at anthesis is achieved during an "**equalization process**":

– The radial movement of the sepals is blocked, while allowing angular movement (the whorl is preserved).

– The distance from the petals to the center of the flower is multiplied by a factor r at each time step (meristem development).

– The contact pressure potential due to all the other primordia is applied to each sepal or petal (Kitazawa and Fujimoto 2015):

$$V_i(R_i, \theta_i) = \sum_{j=1, j\neq i}^{n} \exp(-\frac{d_{ij}}{\delta}) \qquad [3.1]$$

where (R_i, θ_i) are the polar coordinates of organ i among n = 5 organs, and δ is an exponential decay length (different from γ in equation [2.2]).

Primordia descend along the gradient of this potential, which means they move from a position where the potential is high to a point where the potential is lower. To calculate the angular potential gradient, we compared the potential at a virtual primordium i' (R, θ+dθ) and at i (R, θ). The difference in the value of the potential between the real organ and this virtual organ, whose angular position is one degree more, was calculated and used to reposition the n organs by making them go down the potential. From their new positions, we recalculated the gradients of these potentials and obtained new positions, following a recursive process.

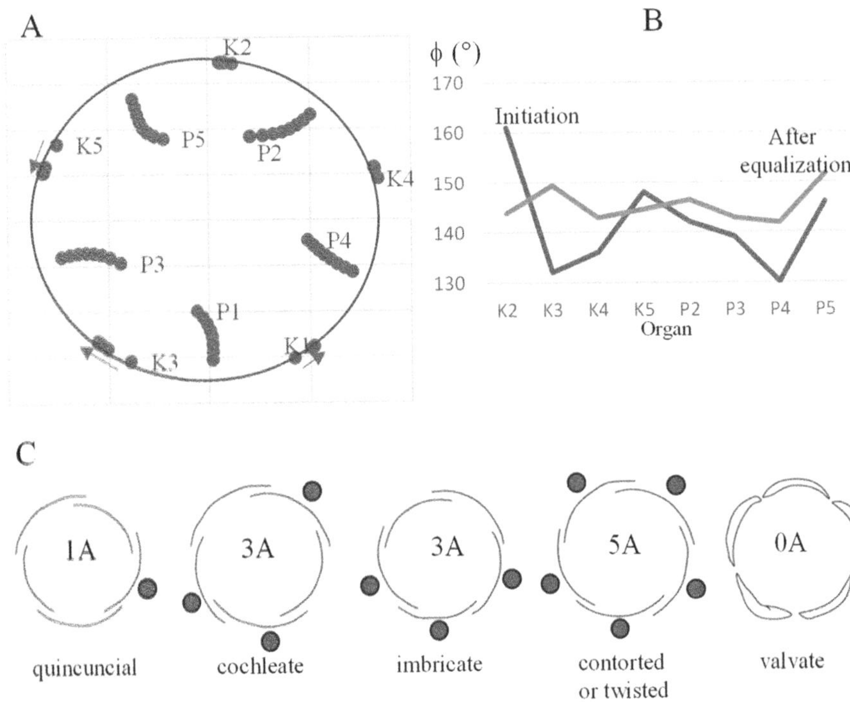

Figure 3.15. *Movement of sepals and petals of Ranunculus repens during the equalization phase according to model [3.1]. (A) The different points represent the evolution of the position of the primordia during this phase. (B) Divergence angles between chronologically consecutive perianth primordia after initiation (Figure 3.14(C)) and after equalization (20 iterations). (C) Aestivations. "xA" inside floral diagrams denote the number of petals overlapping only one of their neighbors (blue dots)*

After the spiral initiation stage, the calyx and corolla primordia are approximatively located on two circles, the center of which is the meristem center. If two primordia contiguous on one circle are closer than average (72° = 360°/5), greater pressure will tend to push them apart. If two others are further apart than average, less pressure will tend to bring them closer together: this is the "equalization process". Ultimately, in a perfect whorled configuration, the potential gradient is zero on both sides of each primordium, the distance from the two neighbors being equal, which ensures the stability of this arrangement.

The contact pressure model explains the transition of *R. Repens* from a spiral to a whorled (actinomorphic) configuration. At the end of the iterative process, the angular distances between the primordia on the whorl are more or less equal.

The frequency of the different aestivations of the terminal and lateral flowers observed by Cunnell (1958) is consistent with the theoretical probabilities for the first three types of aestivations (f = 10/32), as well as for contorted aestivation (2/32). For the extreme lateral flowers, the more the aestivation moves away from the quincuncial type by increasing the number of petals overlapping only one of their neighbors, the lower the frequency.

Anemone

Significant intraspecific variations characterize the genus *Anemone*. Kitazawa and Fujimoto (2018) examined aestivation of five, six and seven (undifferentiated) tepaled flowers in *Anemone* species (Ranunculaceae).

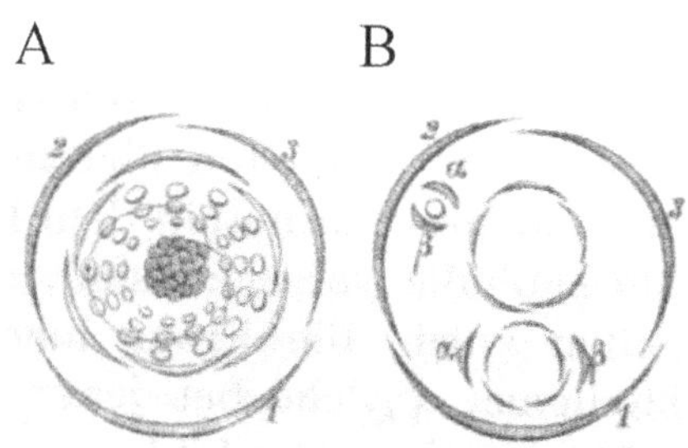

Figure 3.16. *(A) Floral diagram of hexamerous A. nemorosa. (B) Anemone ranunculoides partial inflorescence with three homodromous flowers (Eichler 1878). Involucral leaves are numbered*

The inflorescence is a thyrsoid (thyrse with terminal flower) whose calyx arrangement of the terminal flower is determined by three involucral leaves: the lateral flowers are determined by two bracteoles, the contact pressure exerted by the terminal flower and one of the involucral leaves. Vigor decreases from flower to flower, with the third involucral leaf even being "sterile" (Figure 3.16(C)).

Anemone tomentosa (six tepals)

Each shoot of *Anemone tomentosa* forms a thyrsoid of several flowers. These flowers are preceded by three petiolate involucral leaves in an approximate whorl. Their perianth is composed of five or six tepals (Figure 3.17; Ren et al. 2010).

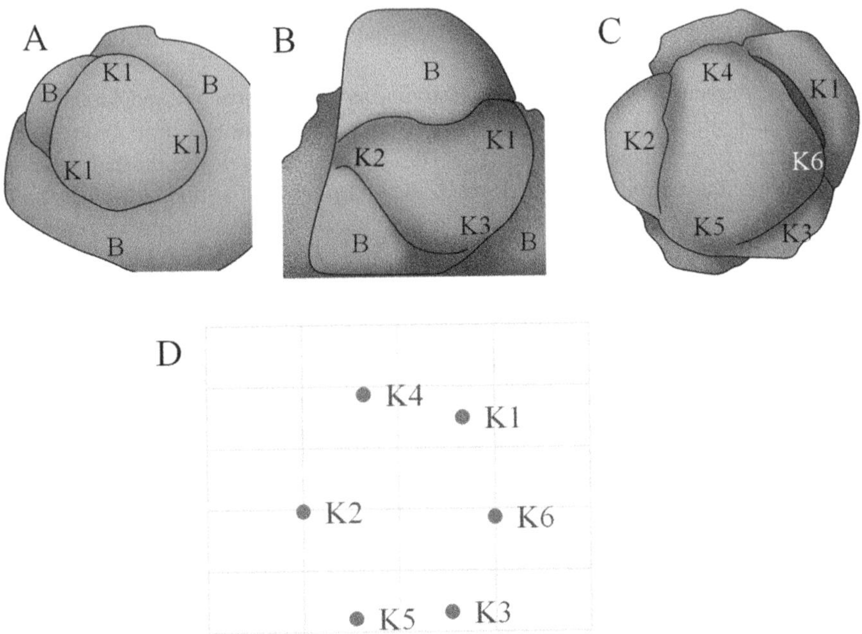

Figure 3.17. *(A) Initiation of Anemone chinensis tepals inside an involucre composed of three connected leaves (Ren et al. 2010). (B and C) Initiation of hexamerous Anemone tomentosa perianth showing the trimerous involucre (Ren et al. 2010). (D) Model. B = involucral leaf*

For modeling, the three involucral leaves are not of ternary symmetry (Figure 3.17(A)): the distances to the center of these three involucral leaves are taken unequal (Table 3.8), which is the origin of a spiral arrangement (non-actinomorphic) of tepals (Figure 3.17(D)).

Organ	Angle (°)	Distance to the center
B1	330	2.5
B2	250	2.3
B3	110	2.1

Table 3.8. *Anemone tomentosa modeling parameters*

Anemone taipaiensis (five tepals)

A shoot of *Anemone taipaiensis* bears one to five flowers that form a thyrsoid. Each flower is preceded by three sessile involucral leaves in an approximate whorl. The five tepals are initiated in a spiral sequence and positioned in an approximate whorl with quincuncial aestivation. There are 60–64 stamens and 60–70 carpels. There is a relatively long plastochron between the initiation of the last sepal and the first stamen. Most of the androecium show irregular spirals with incomplete parastichies. Some flowers show "irregular whorls with sets of nine, 11 or 12 parastichies in both directions". Some flowers are irregularly spiral of mode (8,9), (8, 11), (8, 12), (9, 10), (12, 13), always with additional partial parastichies or single surplus organs (Ren et al. 2010).

The arrangement of the involucral leaves generates a hexamerous arrangement of the first organs, but the sixth of these has a stamen identity (6 = S1 in Figure 3.18(C)). The distances to the center of the three involucral leaves are unequal (Table 3.9), which is the origin of a spiral (non-actinomorphic) arrangement of the tepals.

The observed divergence angles in Figure 3.18(B) are approximatively (K1, K2) $< 180°$; (K2, K3) $> 90°$; (K3, K4) $\approx 180°$; (K4; K5) $> 90°$; $(180 + 90)/2 = 270/2 \approx 135°$ on average. This average value is one of those considered by Kitazawa and Fujimoto (2016a, 2016b, 2018, 2020), but their hypothesis of a truly constant divergence angle between tepals has no empirical basis.

The phyllotaxis of the androecium is irregular because the transition from the small number of perianth organs to the numerous stamens occurs abruptly, without following the path leading to a regular Fibonacci arrangement (see Chapter 2). The number of parastichies depends on the plastochron ratio within the androecium: this must be 1.015 in order to obtain eleven pseudo-parastichies in one direction (Figure 3.18(D); Walch and Blaise 2023).

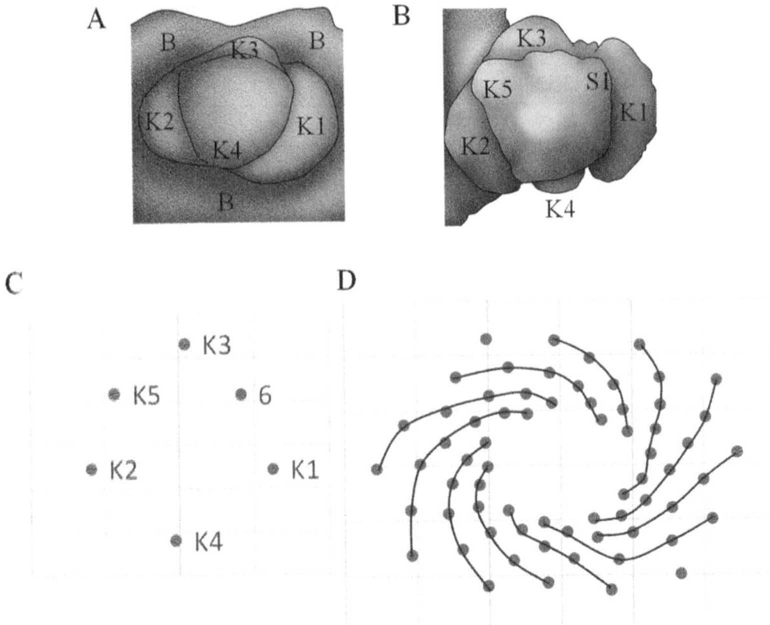

Figure 3.18. *Anemone taipaiensis. (A and B) Perianth initiation according to Ren et al. (2010). Notice the three involucral leaves surrounding the meristem, as well as the sixth organ, which has a stamen identity. (C) Model of perianth. (D) Model of stamen irregular spirals*

Organ	Angle (°)	Distance to the center
B1	50	2.3
B2	130	2.1
B3	270	2
		Plastochron ratio
K1		1.05
K2		1.05
K3		1.05
K4		1.05
K5		1.05
S		1.015

Table 3.9. *Anemone taipaiensis modeling parameters*

According to Schöffel (1932), in *Anemone*, floral phyllotaxis is spiral (Fibonacci or rarely Lucas pattern) in some species and "irregularly whorled" in others. According to Ren et al. (2010), irregularly spiral and irregularly whorled floral phyllotaxis are even present in the same species as *A. taipaiensis*. In the two species with the highest floral organ numbers, *A. tomentosa* and *A. chinensis*, phyllotaxis is "irregularly whorled". The criterion used by Ren et al. (2010) is as follows: phyllotaxis would be whorled when the inclination of the clockwise and counterclockwise parastichies is the same.

Strange phyllotaxis observed by Ren et al. (2010) in *Anemone* are really "quasi-symmetric spirals": according to our model, if plastochron ratios decrease faster than in the canonical sequence (which produces regular spirals), then the chronological sequence of the primordia no longer corresponds to their succession on the ontogenetic spiral with "golden" divergence angles. Then, irregular spiral or quasi-symmetric spirals occur (see Chapter 2).

Anemone hepatica (seven tepals)

Anemone hepatica can have five to 10 tepals. Eichler (1878) presents the diagram of heptamerous and octomerous flowers (Figure 3.19).

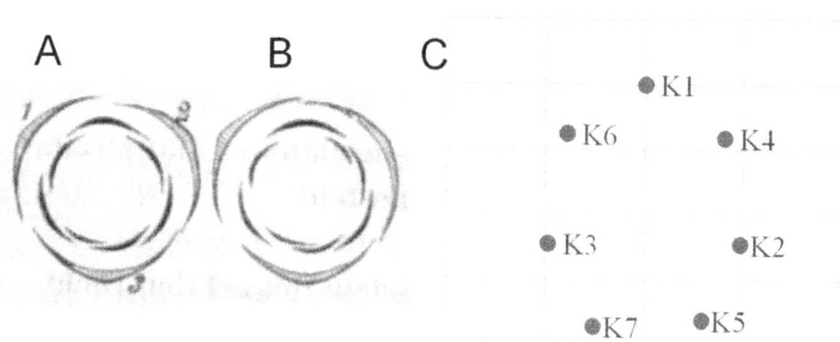

Figure 3.19. *Anemone hepatica: (A) heptamerous and (B) octomerous perianth according to Eichler (1878). Involucral leaves are numbered. (C) Model of heptamerous flower*

Parameters identical to those of a hexamerous flower position the K7 tepal between K2 and K4 in the following of the calyx spiral, as on the Eichler floral diagram. The merism of *Anemone* perianth is not structural but depends on the identity of the first organs.

Organ	Angle (°)	Distance to the center
B1	30	2.5
B2	150	2.5
B3	270	1.5

Table 3.10. *Heptamerous Anemone hepatica modeling parameters. Bn = involucral leaf*

Clematis

The flowers of *Clematis peterae* form thyrsoids. They are preceded by two transverse bracteoles and show four or sometimes five tepals, more than 70 stamens and around 15 carpels. In tetramerous flowers, a first pair of tepals alternates with the two bracteoles, then a second pair alternates with the previous one. "Sometimes one of the sepals in the second pair appears to be divided into two organs (double position) resulting in a pentamerous perianth". A long plastochron separates the tepals from a whorl of four stamens. The small stamens are then quickly initiated in large numbers and form around 15 irregular orthostichies (Ren et al. 2010).

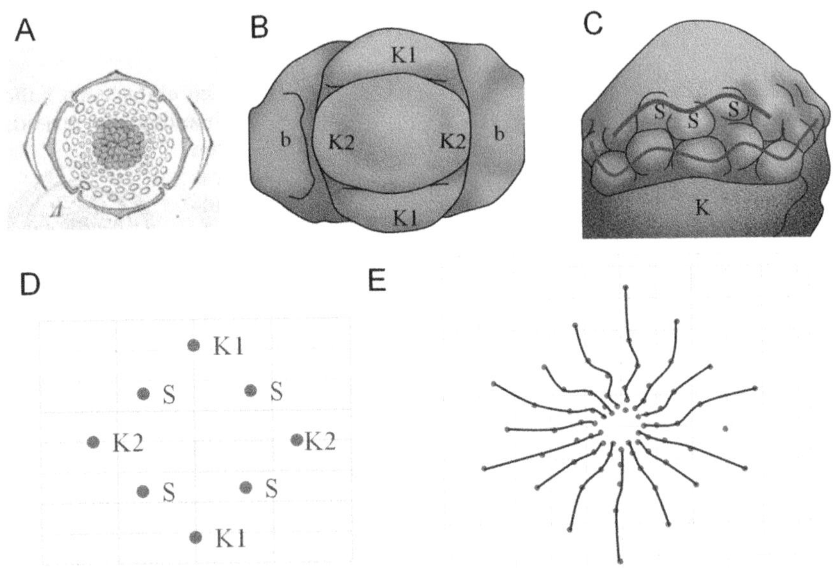

Figure 3.20. *(A) Floral diagram of Clematis integrifolia (Eichler 1878). (B and C) Initiation of tepals and stamens of Clematis peterae, according to Ren et al. (2010). (D) Model of calyx and first stamens position. (E) Model of androecium orthostichies*

Organ	Angle (°)	Distance to the center
b1	0	3
b2	180	3
		Plastochron ratio
K1		1
K2		1.2
K3		1
K4		1.2
S		1.03
S		...

Table 3.11. *Tetramerous Clematis integrifolia modeling parameters*

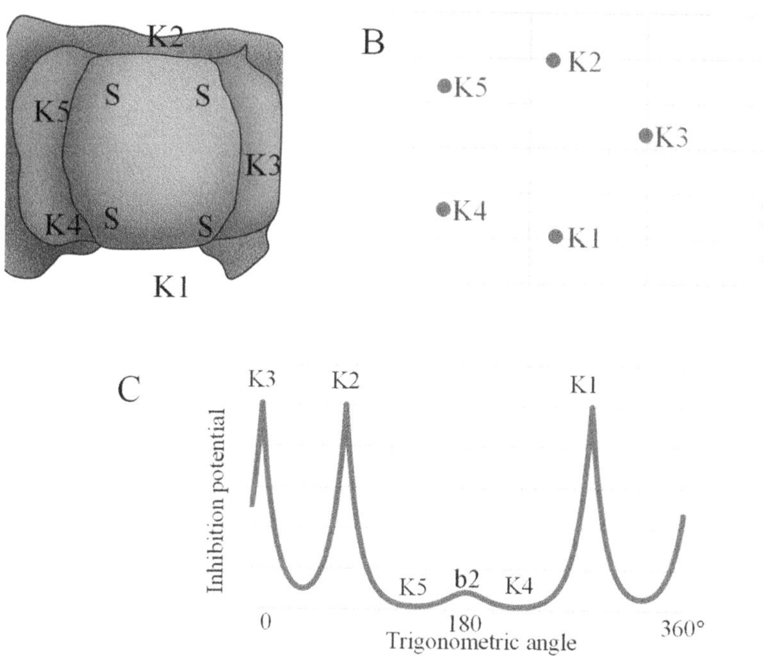

Figure 3.21. *(A) Initiation of the tepals of Clematis peterae, according to Ren et al. (2010) in the pentamerous case. (B) Model. (C) Inhibition profile at initiation of K4*

The transverse positioning of the two bracteoles mechanically generates the dimerous arrangement of the tepals and the first stamens, in perfect symmetry (Figure 3.20(D)). The size of the stamens and the number of pseudo-orthostichies they form depends on the plastochron ratio within the androecium. For r = 1.03, the model presents 15 irregular orthostichies (Figure 3.20(E)).

The calyx of *Clematis peterae* can become pentamerous by "division of the tepals of the second pair" (Ren et al. 2010). The modeling shows that if one of the bracteoles (b2) is sufficiently close to the center of the flower, its inhibition potential creates a "hill" (local maximum) that causes the splitting of the left sepal (Figure 3.21(C)).

Organ	Angle (°)	Distance to the center
b1	0	2
b2	180	1.5

Table 3.12. *Pentamerous Clematis peterae modeling parameters*

4

Symmetry in Lamiales

4.1. Introduction

Actinomorphy is considered to be the ancestral state in Lamiales with *CYC*-like and *RAD*-like genes expressed throughout the floral meristem. Asymmetrical expression with respect to the transverse axis appeared later in Lamiales. We therefore distinguish the early-diverging Lamiales (including Oleaceae), which are generally actinomorphic from the core Lamiales, the vast majority of which are zygomorphic and have an asymmetrical expression with respect to the transverse axis of *CYC* and *RAD* genes in the corolla (Zhong and Kellogg 2015; Zhong et al. 2017) (Figure 4.1).

Independent reversals to radial symmetry from a zygomorphic ancestor in species of core Lamiales (e.g. *Tengia*, *Plantago* and *Mentha*; Zhong et al. 2017) are associated with either an expanded or absent expression of *CYC*-like genes (reviewed in Wessinger and Hileman 2020). Preston et al. (2011) noted that independent shifts to radial symmetry may be linked to similar changes in gene expression, as in *Aragoa* (Plantaginaceae), where radial symmetry is associated with an expanded expression of *CYC2* genes. In *Plantago*, radial symmetry is associated with both an expanded expression of a *CYC2* gene and the loss of another *CYC2* gene and its downstream targets (*RAD* and *DIV*).

We chose plants representative of actinomorphic early-diverging Lamiales (*Olea*, *Ligustrum*), zygomorphic core Lamiales (*Antirrhinum*, *Veronica*, *Salvia*) and plants representative of reversals from zygomorphy to actinomorphy within core Lamiales (*Tengia*, *Bournea*, *Plantago* and *Mentha*) (Figure 4.1).

For a color version of all the figures in this chapter, see www.iste.co.uk/walch/phyllotaxis2.zip.

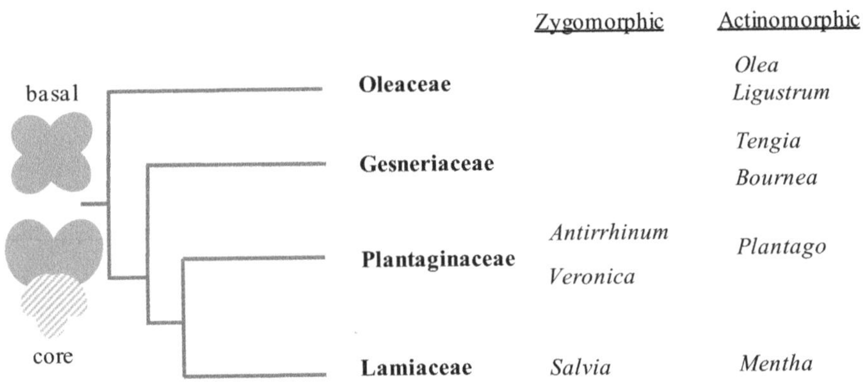

Figure 4.1. *Phylogenetic relationship of the families (and representative species) studied here (from Zhong et al. 2017). Flower outlines indicate the origin of the asymmetrical expression of CYC2 and RAD genes with respect to the transverse axis in the corolla. Floral organs are blue where CYC-like genes are expressed and striped where they are not. The shape of the flower is stylized to represent the symmetry of the corolla at anthesis, as in Hileman (2014)*

Our model was successfully applied to Ranunculales (Chapter 3). However, our tests concerning zygomorphic Lamiales required a modification of the basic model. Indeed, zygomorphy is characterized by the symmetrical initiation of organs with respect to the dorsoventral axis. The necessary modification consists, in addition to the initiation of any organ (located outside the dorsoventral axis), of initiating another primordium, which is symmetrical to it with respect to the dorsoventral axis. These two primordia must be initiated simultaneously; otherwise, the first would have an influence on the position of the second. We call this process "**twinning**".

4.2. Oleaceae

Oleaceae belongs to the early-diverging Lamiales, along with Plocospermataceae, Carlemanniaceae and Tetrachondraceae. Their thyrsoid inflorescences form dichasia with opposite bracteoles.

Olea

The inflorescence apex (A) of *Olea europaea* (Oleaceae) produces decussately arranged bracts (e.g. B on Figure 4.2(A)), in the axil of which second-order flowers (e.g. F2) are initiated. These second-order branches produce decussately arranged second order bracteoles (a and b) (Naghiloo et al. 2013) (Figure 4.2(A)). According to Eichler (1875), the first pair of sepals in *O. europaea* is dorsoventral, and the second pair is transverse (2 + 2 opposite-decussate arrangement) (Figure 4.2(B)). To produce this pattern, the inhibition generated by bracteoles a and b must be stronger than that generated by the two dorsoventral sources (Table 4.1: Ra = Rb = 2; Rad = RB = 3): therefore, K1 and K2 are located on the dorsoventral axis, that is, as far away as possible from a and b.

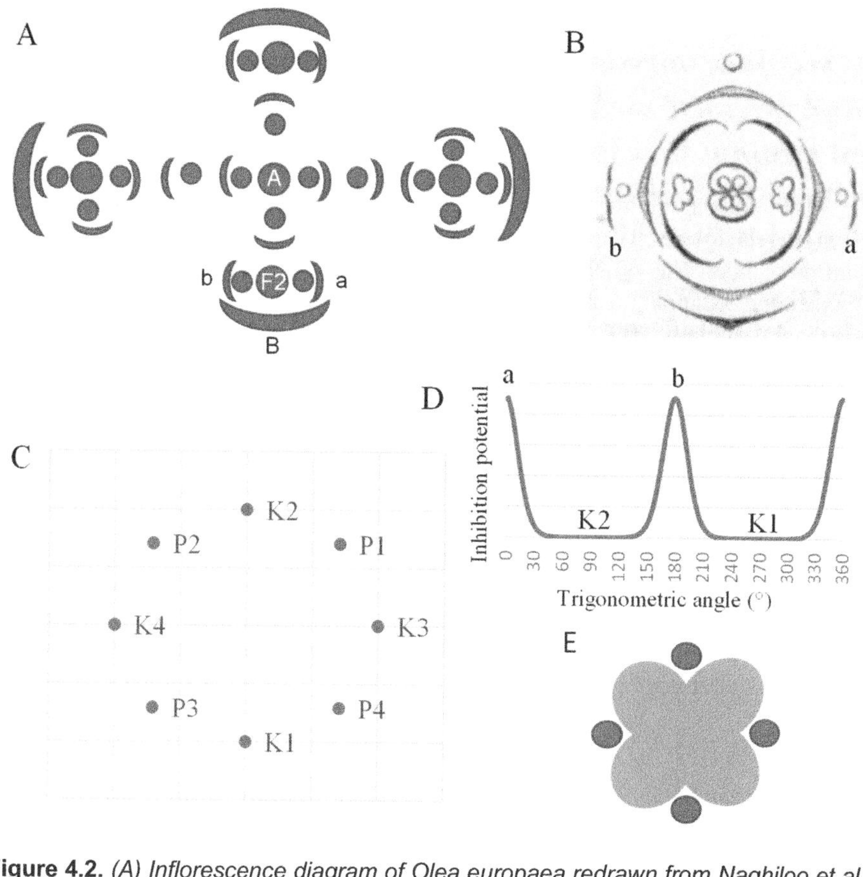

Figure 4.2. *(A) Inflorescence diagram of Olea europaea redrawn from Naghiloo et al. (2013). A = apex; a, b = bracteoles and B = subtending bract of F2. (B) Floral diagram from Eichler (1875). (C) Model of the O. europaea perianth. K = sepal, P = petal, numbers reflect the order of initiation. (D) Inhibition potential profile at the initiation of the first sepals (K1, K2): the angular locations of existing maxima are those of bracteoles a (0°) and b (180°) in (A). Simultaneous initiation of K1 and K2 takes place at the equal potential minima (270° and 90°). (E) This is the expression of the single CYC2-like gene (in blue) in the corolla of Ligustrum lucidum (Oleaceae) (based on Zhong and Kellogg 2015). The shape of the flower is stylized to represent the symmetry of the corolla at anthesis. Green discs indicate sepal location*

Organ	Angle (°)	Distance to the center
Adaxial source	90	3
Bract	270	3
a	0	2
b	180	2

Table 4.1. *Parameters used to model the perianth of Olea europaea. The two bracteoles (a and b) are angularly ($\theta a = 0°$ and $\theta b = 180°$) and radially ($Ra = Rb = 2$) symmetrical*

The adaxial source for non-terminal flowers (e.g. F2: Figure 4.2(A)) is considered to be the contact pressure from the bract located above in the inflorescence. The process of initiating a whorl is very different from that of generating a spiral (Walch and Blaise 2022a): the initial conditions have to be perfectly symmetrical. Therefore, when the first primordium of the first whorl is initiated (e.g. K1; Figure 4.2(D)), the inhibition potential exhibits two minima with equal inhibition potentials due to the perfect symmetry of these initial conditions (thus, Curie's principle is verified: the cause being bracteole symmetry, the effect being the symmetry of the first sepals). Therefore, two primordia will be initiated simultaneously at these minima (K1 and K2: if it is not simultaneous, the inhibition produced by K1 will disrupt the position of K2). These primordia will be symmetrical (a new cause), which propagates the symmetry to the transverse sepals (K3 and K4; the effect of this cause), and so on, to the corolla which forms a tetramerous whorl alternating with the calyx (Figure 4.2(B)). This alternation continues with the single pair of stamens and then the two carpels (not shown). The flower is therefore polysymmetric.

Ligustrum

Ligustrum lucidum (Oleaceae) has a tetramerous actinomorphic perianth such as *O. europaea.* Zhong and Kellogg (2015) showed using RNA in situ hybridization that in *L. lucidum*, *CYC2*-like and *RAD*-like genes are expressed throughout the corolla (Figure 4.2(E)). They concluded from this broad expression pattern in early-diverging Lamiales that asymmetrical expression in adaxial petals appeared later in the core Lamiales (Figure 4.1). The symmetry of the causes – here, *CYC*-like gene expression – produces an effect – the arrangement of the perianth organs – of the same symmetry (Curie's principle): asymmetry with respect to the transverse axis in *Antirrhinum* (Figure 1.12(B)) and symmetry in *Ligustrum*.

4.3. Gesneriaceae

Gesneriaceae is a species-rich (ca. 3,400 species), predominantly zygomorphic family (e.g. *Haberlea fernandi-coburgii*, Figure 4.3(B)), in which several independent reversals to actinomorphy have occurred (e.g. *Ramonda myconi*, Figure 4.3(A); *Tengia scopulorum*, Figure 4.3(E); *Bournea leiophylla*, Figure 4.3(G)).

Tengia (syn. Petrocodon scopulorum)

Tengia scopulorum (Gesneriaceae) has been called a "natural peloria" because it exhibits an almost perfect actinomorphic flower at anthesis. The calyx and corolla are pentamerous (Pang et al. 2010).

Pang et al. (2010) showed using RNA in situ hybridization that *TsCYC1C* and *TsCYC1D* have similar expression patterns with strong signals detected in all five petals; whereas, *TsCYC2A* and *TsCYC2B* are only transiently expressed in the very early floral meristem. Their results suggest that the expansion of *TsCYC1C* and *TsCYC1D* expression from the dorsal to the ventral petals (Figure 4.3(C)) is likely to be responsible for the evolution from zygomorphy to a fully dorsalized actinomorphic corolla in this species (Figure 4.3(F)).

The initiation of sepals and petals in *Tengia* is bilaterally symmetrical, the shape of the flower being elongated along the dorsoventral axis (**irregular pentagon**; Figure 4.3(C)). At anthesis, the corolla becomes perfectly actinomorphic with a quasi-circular shape resulting in a whorled arrangement of these organs (**regular pentagon**; Figure 4.3(E)). At initiation, in Figure 4.3(C), the angle whose vertex is the abaxial petal is shown in blue. Figure 4.3(D) shows that this angle is smaller than an angle whose vertex is a lateral petal. Figure 4.3(E) shows that after development, this angle became larger to the detriment of the angles whose vertices are other petals.

The transition from a monosymmetrical initiation to an actinomorphic (whorled) arrangement is achieved during an "**equalization**" process that we modeled (see Figure 3.15): on the same whorl, the contact pressure between the different primordia tends to equalize the angles between pairs of contiguous primordia. If two of them are closer than would be expected by equal spacing (72° = 360°/5), greater pressure will tend to push them apart. If two primordia are further apart than equal spacing, less pressure will tend to bring them closer together. Therefore, at the end of this process, the perianth of *Tengia* changes from monosymmetric to radially symmetric.

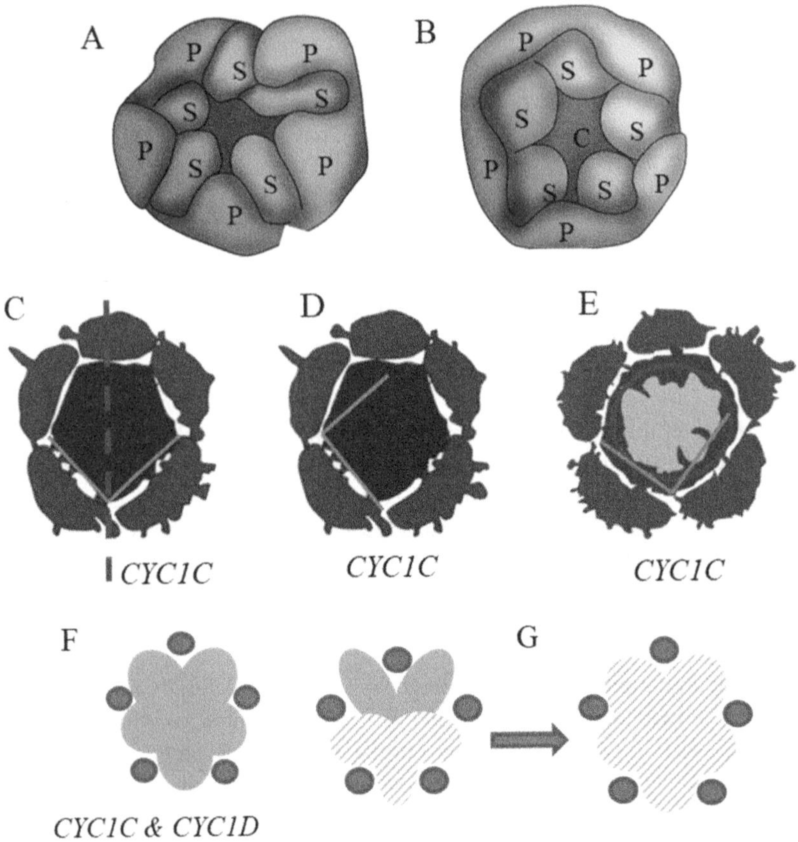

Figure 4.3. *Flowers before anthesis. (A) Pentamerous actinomorphic (Ramonda myconi) and (B) zygomorphic (Haberlea fernandi-coburgii) flowers in Gesneriaceae (redrawn from Citerne et al. 2000). (C–E) Expression of TsCYC1C in floral meristems in Tengia scopulorum, from Pang et al. (2010): (C–D) TsCYC1C (brown) is detected in sepals, all five petal and stamen primordia, with the measurement of the angle whose vertex is the abaxial petal before equalization in blue. (E) TsCYC1C expression becomes gradually weaker in stamens (yellow), but the signal is strong in sepals and petals when stamens begin to enlarge. Blue: The measurement of the angle whose vertex is the abaxial petal before equalization. (F) Expansion of TsCYC1C and TsCYC1D expression of CYC-like gene in the corolla of Tengia (in blue). Green discs indicate sepal location. (G) Changes in the expression of a CYC-like gene during corolla development in Bournea leiophylla (Zhou et al. 2008). Floral organs are blue where CYC-like genes are expressed and striped where they are not. The shape of the flower is stylized to represent the symmetry of the corolla at anthesis. Green discs indicate sepal location. Floral symmetry in Bournea shifts from zygomorphy to actinomorphy during development*

Bournea (syn. Oreocharis leiophylla)

In *Bournea leiophylla* (Gesneriaceae), zygomorphy is observed in the early stages of floral development; however, flowers are actinomorphic at anthesis. Zhou et al. (2008) RT-PCR results detected no *BlCYC2* transcripts in sepals, petals and stamens in all stages of flowers that they tested, suggesting that *BlCYC2* may not play a significant role in floral development in *B. leiophylla*. They used RNA in situ hybridization that specifically detected the *BlCYC1* gene on the dorsal side of the floral apex during early development. At later stages, *BlCYC1* transcripts are barely detectable in the flower (Figure 4.3(G)). The signal of *BlRAD* matches that of *BlCYC1*. This shift is in accordance with Curie's principle: asymmetric dorsoventral expression is the cause of early zygomorphy, while radially symmetrical expression causes late actinomorphy.

4.4. Plantaginaceae

Antirrhinum

Bravais and Bravais (1837) studied the shift from decussate to spiral arrangement during the transition from the vegetative to the reproductive stage in selected species of Lamiales (*Antirrhinum*, *Verbena* and *Veronica*). The inflorescence of snapdragon, *Antirrhinum majus*, is a raceme with an apical meristem that does not form a terminal flower and produces floral meristems following a (3, 5) Fibonacci pattern (Figure 4.4(A): Carpenter et al. 1995).

Bull-Hereñu et al. (2022) noted that, in racemose inflorescences, rigid bracts compress the young, developing flowers. When these bracts exert extensive pressure, floral primordia become squeezed and even flattened (FM14 in Figure 4.4(A)).

We hypothesized that in the absence of bracteoles, which is the case in *A. majus*, the forces shaping the perianth are not due to biochemical activity but to contact pressure from the bracts subtending the flowers above each developing flower in the inflorescence (Walch and Blaise 2022b).

Antirrhinum majus flowers have five sepals, five petals, five stamens and two carpels (Figure 4.4(B)). Corolla aestivation in mature flowers is "antirrhinoid" (Eichler 1875) or descending cochlear, with the dorsal lobes being external in bud (P1–P2: Figure 4.4(C)), as is also found in *Globularia* and *Gratiola* (Plantaginaceae; Bello et al. 2004).

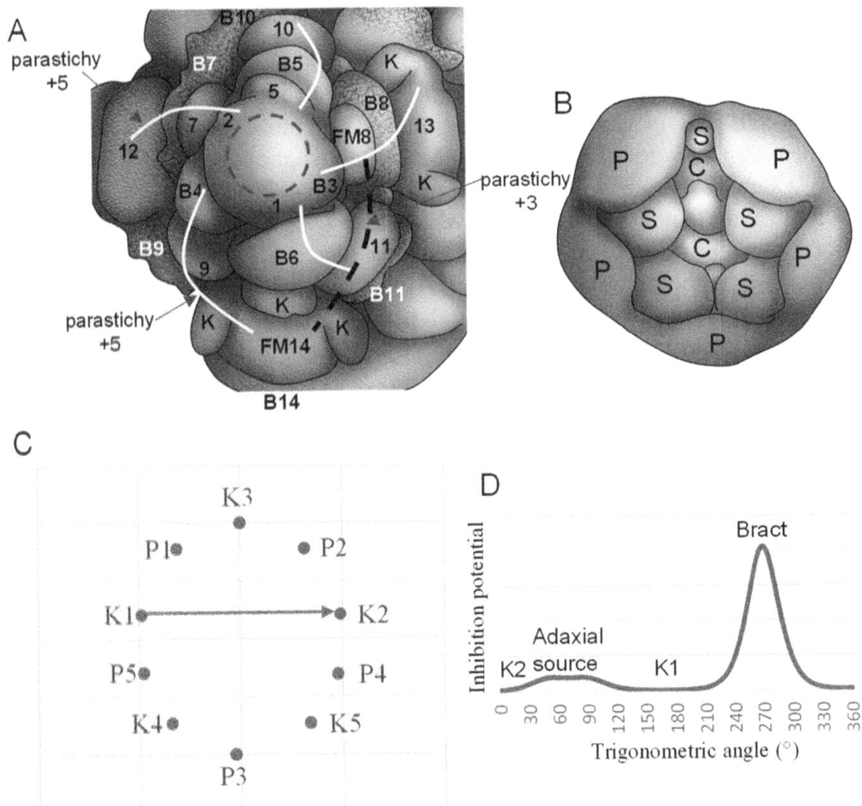

Figure 4.4. *(A) The inflorescence apex of Antirrhinum majus (redrawn from an SEM from Carpenter et al. 1995). Numbers correspond to floral meristems: 1 being the youngest; Bn corresponds to the bract number; bracts B7-B10 are removed to make the floral meristems visible. In white lines, the five left parastichies (+5 between successive primordia on the parastichy); dark dashed line: one of the three right parastichies (+3 between successive primordia on the parastichy). (B) The floral meristem at stage 6 redrawn from Vincent and Coen (2004). (C) Model of the floral meristem. Arrows indicate pairs of primordia initiated simultaneously as bilateral twins. (D) Profile of the inhibition potential at the initiation of K1. K = sepal, P = petal, S = stamen, C = carpel*

The sepals are united in their base to form a corolla tube ending in five separate lobes. The different petals (dorsal, lateral and ventral) have different shapes and sizes. The dorsal petals have large lobes; whereas, the lateral and ventral petals have smaller lobes. The corolla is asymmetric with respect to the transverse axis, which gives the flower its strong zygomorphic shape at anthesis (Figure 1.1(C)).

Organ	Angle (°)	Distance to the center
Adaxial source (B_{n-8})	90	2
Bract (B_n)	270	1.5
B_{n-3}	50	2
B_{n-5}	120	3

Table 4.2. *Parameters used to model the flower of Antirrhinum majus. B_{n-3} and B_{n-5} are the bracts located on the same two parastichies as the developing floral meristem (FMn) and the adaxial source B_{n-8} is on the same orthostichy. Angle (θ) and distance from the center (R) are the polar coordinates of the organ*

In the absence of bracteoles, we hypothesized that four sources of inhibition influence the initiation of sepals in *A. majus* (Table 4.2): the bract subtending the flower (e.g. B14 subtending FM14; $\theta_B = 270°$ = absolute angle of bract B), the two bracts located above the flower on the two parastichies passing through this flower (e.g. $\theta_{B11} = 50°$; $\theta_{B9} = 120°$) and the large adaxial bract located on the pseudo-orthostichy passing through the flower (e.g. B6). The chosen inclination of these three upper bracts is based on the SEM of Carpenter et al. (1995) (Figure 4.4(A)). The sepal initiation sequence generated by the model (Figure 4.4(C)) matches that of FM14.

Luo et al. (1996, 1999) found that the floral meristems of *Antirrhinum majus* from stage 2 onwards show only dorsal expression of both *CYC* and *DICH* (Figure 1.12(B)); this expression pattern confers symmetry with respect to the dorsoventral axis but asymmetry with respect to the transverse axis, in agreement with Curie's principle.

According to our model, contact pressure from the bracts subtending the flowers above each developing flower in the inflorescence is the cause of the arrangement of the calyx. As the two parastichies (+3 and +5) are asymmetrical with respect to the dorsoventral axis of the flower Fn, the sepals should also be asymmetrical according to Curie's principle. We hypothesize that *CYC*-like genes force the bilateral initiation of the organs (pairs of primordia are initiated simultaneously as bilateral twins) (arrows in Figure 4.4(C)). This requires that *CYC*-like genes be expressed at the initiation of the perianth. Compared to *Tengia*, we may also think that asymmetric *CYC/DICH* gene expression prevents an equalization towards actinomorphy before anthesis, the perianth remaining monosymmetrical.

Plantago

The inflorescence of *Plantago major* is a dense spike (a raceme with sessile flowers on an elongated axis), the flowers being initiated in a Fibonacci spiral that can reach up to mode (13, 21) as already noted by Braun (1831). The initiation of the two ventral sepal primordia is followed by that of the two dorsal primordia. The four stamens emerge simultaneously shortly before petal initiation. The emergence of the petal primordia is unidirectional from the abaxial to the adaxial side. There is no evidence that the adaxial petal is formed by the fusion of two petal primordia. By mid-development, the corolla is actinomorphic (Bello et al. 2004) (Figure 4.5(C)).

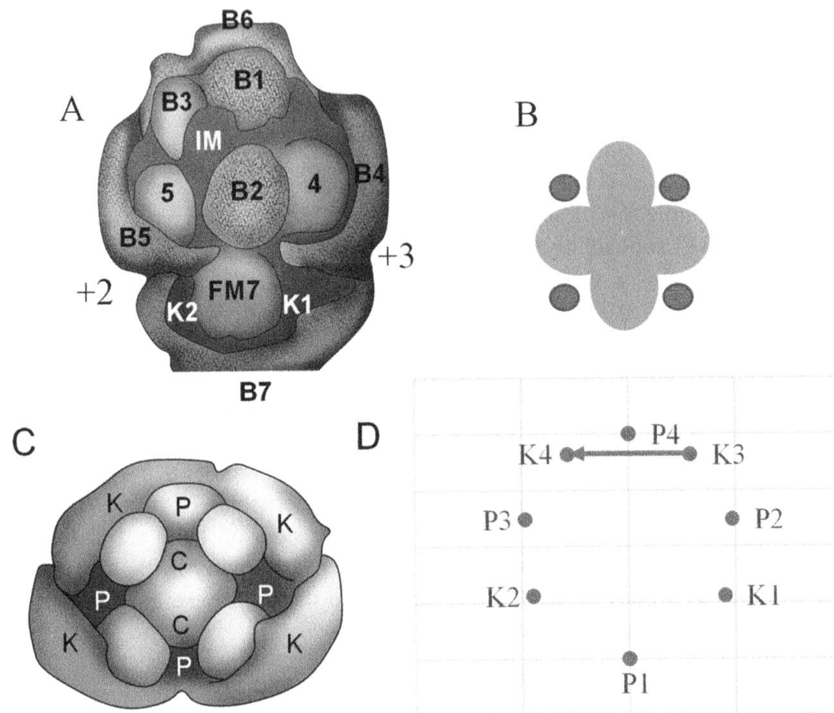

Figure 4.5. *Plantago major. (A) Inflorescence meristem redrawn from Bello et al. 2004. IM = inflorescence meristem apex; numbers = floral meristems; B = bracts; K1 and K2 = first sepal primordia of FM7. (B) Expression of the CYC2-like gene in the corolla, according to Preston et al. (2011). Petals are blue where CYC-like genes are expressed. The shape of the flower is stylized to represent the symmetry of the corolla at anthesis. Green discs indicate sepal location. (C) Floral meristem (redrawn from Endress 1998). (D) Model of the initiation pattern of the perianth. Arrow = organ twinning*

Organ	Angle (°)	Distance to the center
Adaxial source (B2)	90	2
Bract (B7)	270	2.5
a (B4)	40	2.5
b (B5)	135	2

Table 4.3. *Parameters used to model the flower of P. major. Angle (θ) and distance from the center (R) are the polar coordinates of the bracts (numbers refer to the bracts surrounding FM7)*

Floral organ initiation begins after the formation of floral meristem FM7 (Figure 4.5(A)), well before the initiation of floral organs in *Antirrhinum* (Figure 4.4(A)). The mode of the Fibonacci spirals is (2, 3). On the SEM of the inflorescence apex, bract B5 (parastichy +2) is closer to FM7 than bract B4 (parastichy +3).

Eichler (1875) noted that in *Plantago*, the bracteoles are always "Unterdrückt" (cryptic). He included "Unterdrückt" organs, which do not form visible structures, to his diagrams from mainly theoretical assumptions (Eichler 1875, p. 225; discussed in Choob 2022). These are represented by dashed patterns (e.g. Figure 4.7(B)). According to our interpretation, the contact pressure from the bracts subtending the flowers above each developing flower in the inflorescence corresponds to these cryptic bracteoles.

Preston et al. (2011) showed that in *P. major*, the single *CYC*-like gene is expressed throughout the floral meristem during early development, and during mid- to late-stage development, in all organs of the flower (i.e. in the dorsal, lateral and ventral domains) (Figure 4.5(B)). Like in *Antirrhinum*, the position of the two bracts (which are a source of contact pressure) are asymmetrical with respect to the dorsoventral axis. This should be reflected in the perianth, but due to the ubiquitous expression of the *CYC*-like gene, these organs are twinned, which ensures the actinomorphy of the perianth.

Veronica

The inflorescence of *Veronica* is a raceme with spirally arranged buds (corresponding to a Fibonacci spiral), like that of *Antirrhinum* and *Plantago* (Bull-Hereñu and Claßen-Bockhoff 2011). The calyx of *Veronica* is most often a tetramerous St. Andrew's cross, but some *Veronica* species have five sepals: for example, *V. persica* (occasionally) and *V. austriaca subsp. teucrium* (almost always), whose fifth sepal is much smaller than the others (Knuth 1899). Corolla aestivation is "rhinanthoid", with ventral lobes (P1–P2–P3) external in bud. Tetramerism of the corolla at anthesis is the result of the fusion of two dorsal petals, which makes the upper lobe the largest of the four (Eichler 1875). The flower is thus zygomorphic (Figures 4.6(A) and (C)).

In *Veronica*, the abaxial pair of calyx lobes is formed first and continues to be larger than the adaxial pair throughout development. Stamen primordia are already initiated and quite large when the corolla lobes begin to form in an abaxial to adaxial direction (Kampny et al. 1993).

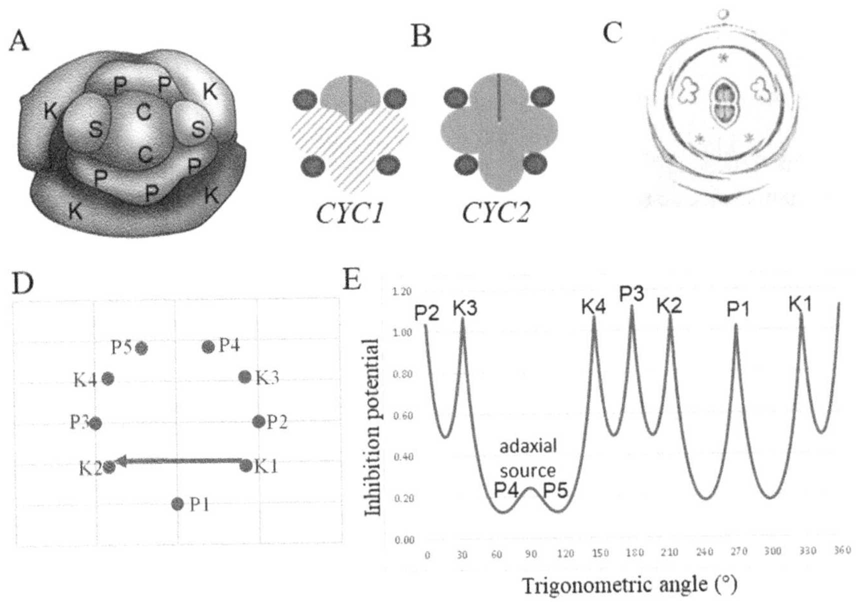

Figure 4.6. *(A) Floral development of Veronica spicata (redrawn from Endress 1999). (B) Expression of VmCYC1 and VmCYC2 in the corolla of V. montana (redrawn from Preston et al. 2009). Floral organs are blue where CYC-like genes are expressed and striped where they are not. The shape of the flower is stylized to represent the symmetry of the corolla at anthesis. Green discs indicate sepal location. (C) Floral diagram of V. chamaedrys, according to Eichler (1875). (D) Model of V. chamaedrys. Arrow = organ twinning. (E) Profile of the inhibition potential at the initiation of P4/P5*

The expression of *CYC*-like and *RAD*-like genes was investigated in *V. montana* by reverse transcriptase polymerase chain reaction (RT-PCR) and RNA in situ hybridization by Preston et al. (2009). They found that *VmCYC1* and *RAD*-like genes are expressed exclusively in the dorsal region of the floral meristem. By contrast, *VmCYC2* is expressed in both vegetative and floral tissues and is most abundant in all regions of the petals (Figure 4.6(B)).

Organ	Angle (°)	Distance to the center
Adaxial source	90	1.3
Bract	270	2.5
a	40	3
b	135	2.5

Table 4.4. *Parameters used to model the flower of Veronica chamaedrys. Angle (θ) and distance from the center (R) are the polar coordinates of the organ. Bract angles are the same as in Plantago*

Although the inflorescence of *Veronica* is a raceme with spirally arranged buds, like that of *Plantago*, we had to reinforce the inhibition from the adaxial source (compare Tables 4.3 and 4.4), which Eichler intuited by adding a cryptic adaxial bracteole (Figure 4.6(C)). This creates an adaxial local maximum at $\theta = 90°$, which leads to a "dédoublement" of the adaxial petal (P4 and P5), as opposed to the single adaxial petal in *Plantago* (Figure 4.6(E)). Like in *Plantago*, however, ubiquitous *CYC2*-like gene expression leads to the twinning of (K1, K2) primordia at their inception.

4.5. Lamiaceae

Salvia

The inflorescence of *Salvia verticillata* is a verticillaster: a thyrse with opposite-decussate bracts, each axillating a dichasium of **scorpioid cymes** (Figure 4.7(A)). In *Salvia*, bracteoles are not visible, and they are "cryptic" (Eichler 1875: dashed bracteoles) (Figure 4.7(B)). The calyx is bilabiate and gamosepalous; sepals K1–K3–K2 and K4–K5 fuse during development (Figures 4.7(A–D)) (Claßen-Bockhoff and Bull-Hereñu 2013; Naghiloo 2020). The flower of *Salvia* is strongly zygomorphic. The corolla of *S. officinalis* is antirrhinoid (descending cochlear) with dorsal P1 and P2 (Figure 4.7(B)).

The angular coordinates of the cryptic bracteoles (Table 4.5: a and b), corresponding to the point of contact pressure exerted by the two neighboring floral meristems (F3' and F4 on F3; Figure 4.7(A)), were chosen to correspond to the SEM of the *S. verticillata* inflorescence (from Claßen-Bockhoff and Bull-Hereñu 2013).

The points of contact pressure are asymmetrical (white line on Figure 4.7(A)): in the model, it is thus necessary to force the bilateral initiation of the organs of the corolla by twinning the primordia.

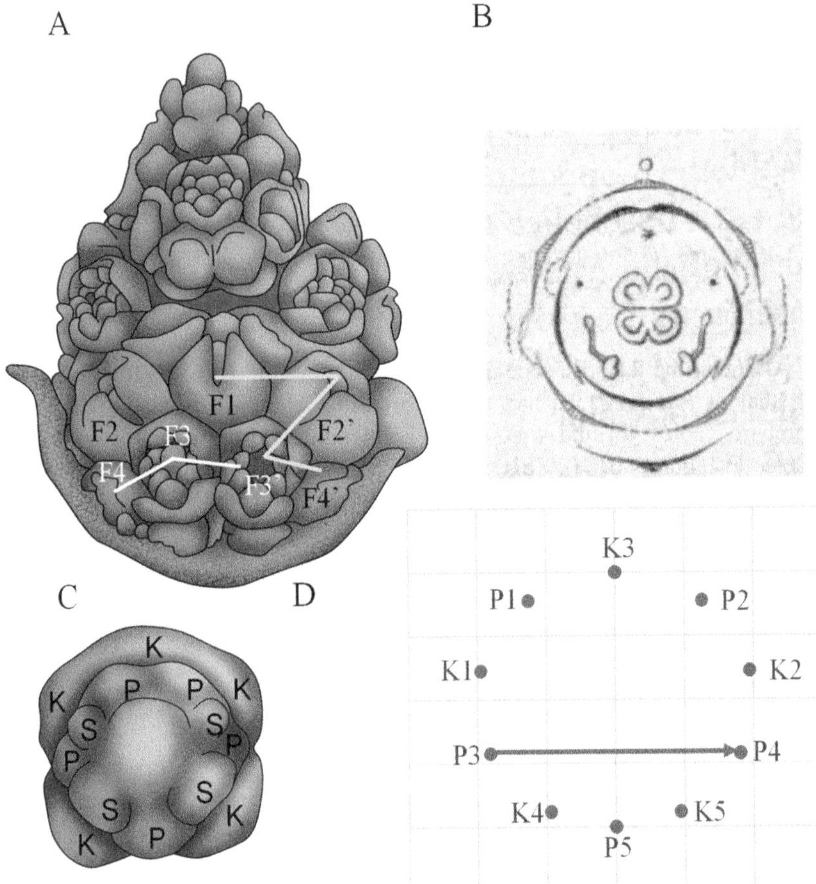

Figure 4.7. *(A) The terminal thyrse of Salvia verticillata (redrawn from Claßen-Bockhoff and Bull-Hereñu 2013) with scorpioid cymes (F2-F3-F4 and F2'-F3'-F4' in yellow lines). White lines show the direction of pressure forces applied to the F3 gravity center. (B) Floral diagram of S. officinalis (Eichler 1875). (C) SEM of the floral meristem of S. nemorosa redrawn from Naghiloo (2020). (D) Modeling of S. officinalis. Arrow = organ twinning*

Organ	Angle (°)	Distance to the center
Adaxial source	90	2.5
Bract	270	2
b	200	3
a	360	4

Table 4.5. *Parameters used to model the flower of Salvia officinalis. Angle (θ) and distance from the center (R) are the polar coordinates of the organ. Angles for the direction of the pressure forces are taken from Figure 4.7(A)*

Mentha

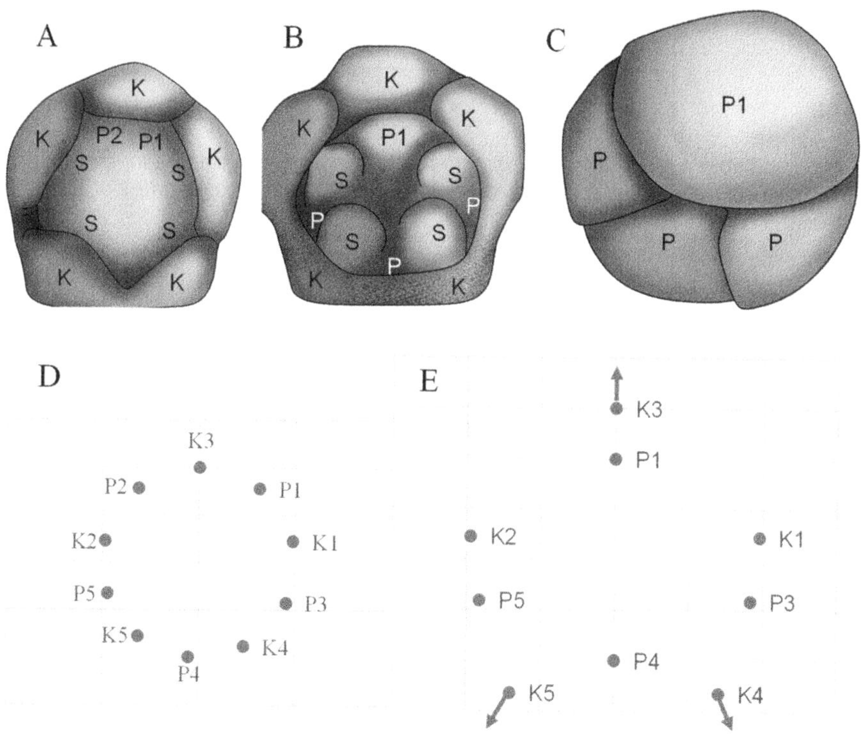

Figure 4.8. *Floral development of Mentha longifolia: (A) The flower is initially pentamerous and bilaterally symmetrical. (B) The two dorsal petals fuse, and (C) the corolla becomes tetramerous and almost radially symmetrical (redrawn from the SEMs of Zhong et al. (2017); the same scale is used for the three drawings). Model of the calyx and corolla initiation pattern in M. longifolia (D) before and (E) after displacing K3, K4 and K5 outwards. Arrows indicate the direction of growth of the meristem*

Like in many Lamiaceae, the inflorescence of *Mentha* is a verticillaster. The calyx of the mature flower is composed of one upper three-toothed lip and one lower two-toothed lip. The corolla is pentamerous at initiation and becomes tetramerous during development. The two dorsal petals are initiated in alternation with the dorsal sepals (Figure 4.8(A)) but quickly fuse into a single adaxial petal (Figure 4.8(B)). Later in development, the adaxial petal develops faster than the other petals and the corolla becomes four-lobed, the upper lobe being wider than the others (Figure 4.8(C)) (Zhong et al. 2017).

Organ	Angle (°)	Distance to the center
Adaxial source	90	2.5
Bract	270	2
b	200	3
a	340	4

Table 4.6. *Parameters used to model the flower of Mentha longifolia. Angle (θ) and distance from the center (R) are the polar coordinates of the organ*

To model the floral development in *M. Longifolia*, sepals K3, K4 and K5 must be displaced outwards; this move allows the fusion of petals P1 and P2, so that the adaxial petal is opposite K3. Indeed, if K3 were closer to the center, its inhibition would prevent the development of a purely adaxial petal (Figures 4.8(D–E)).

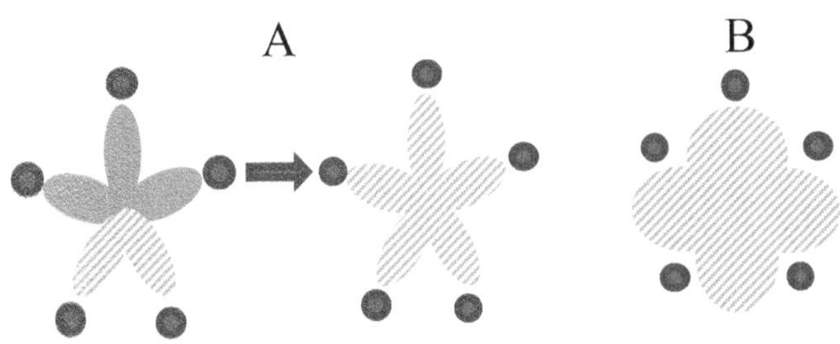

Figure 4.9. *Floral development of Mentha longifolia: CYC2B1 gene expression in (A) the calyx and (B) the corolla of M. longifolia (Zhong et al. 2017). Blue = expression of CYC2-like genes in the corolla; striped indicates lack of expression; green discs indicate sepal location*

Zhong et al. (2017) investigated the expression of *CYC*-like genes in *M. longifolia* using RNA in situ hybridization. During the early stages of development, *MlCYC2B1* was detected in sepals, but this expression is only transient (Figure 4.9): "finally, loss of *MlCYC2A* and *MlRAD1* and altered/transient expression of *MlCYC2Bs* might account for mature radial symmetry in *Mentha*" (Zhong et al. 2017).

5

Constitutional versus Positional Monosymmetry

5.1. Introduction

According to Endress (1999), there are two categories of zygomorphy:

1) "Constitutional zygomorphy" with elaborate monosymmetric flowers (e.g. Lamiales). Predominantly zygomorphic clades, where almost all members have bilaterally symmetrical flowers, have this category of zygomorphy.

2) Taxa with monosymmetric flowers also occur in groups with predominantly polysymmetric flowers. In some cases, monosymmetry arises by late ontogenetic modifications. According to Vöchting (1885), this monosymmetry could be induced by gravity. This category of zygomorphy is called "positional zygomorphy".

For example, in zygomorphic Ranunculaceae (Figure 5.1), zygomorphy is established late in development (after organogenesis is complete) compared to many other zygomorphic eudicot species (e.g. within the order Lamiales) (Jabbour et al. 2009).

In the light of these two categories of zygomorphy, Endress (1999) asks the following questions: is there one mode or different modes that control shifts in floral symmetry in Lamiales and in angiosperms in general? What is the genetic difference between constitutional and positional zygomorphy? We now have the elements to answer these questions.

For a color version of all the figures in this chapter, see www.iste.co.uk/walch/phyllotaxis2.zip.

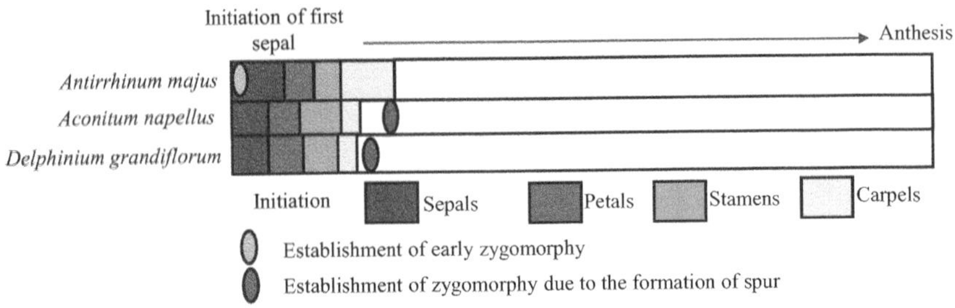

Figure 5.1. *The timing of events during floral development in two species of Ranunculaceae (Aconitum napellus and Delphinium grandiflorum) and in one core eudicot species (Antirrhinum majus). The relative time axis indicates the percentage of development already accomplished (calculated as the ratio of the floral bud base width to the floral base width of the adult flower) (from Jabbour et al. (2009))*

5.2. Saxifragales

Saxifraga

Most *Saxifraga* species, such as *S. granulata*, are actinomorphic. Calyx arrangement is quincuncial and petal initiation is simultaneous (Figure 5.2).

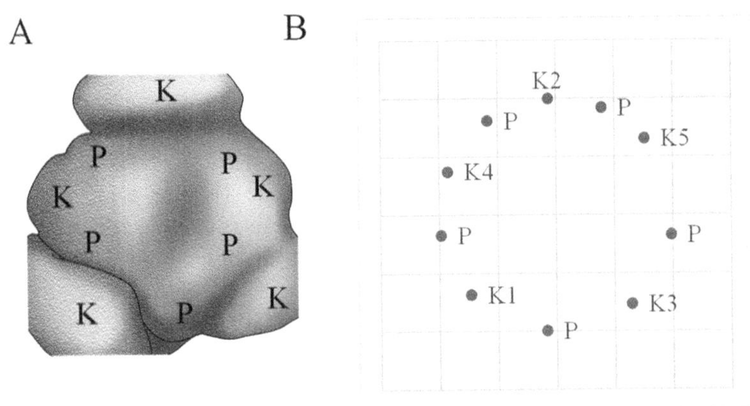

Figure 5.2. *(A) Development of S. fragilis according to Soltis and Hufford (2002). (B) Model of the arrangement of perianth organs in an actinomorphic Saxifraga. The petals are not numbered because their initiation is simultaneous*

Organ	Angle (°)	Distance from the center
Adaxial source	90	3
Bract	270	2.4
β	10	1.6
α	160	2.1

Table 5.1. *Modeling parameters for Saxifraga fragilis*

By contrast, *Saxifraga stolonifera* is an oblique zygomorph. Although its calyx is quincuncial and petal initiation is similar to that of actinomorphic *Saxifraga*, the growth of petals P1 and P2 symmetrical with respect to an oblique axis (36° relative to the dorsoventral axis) is faster than that of the other three petals, so that the corolla becomes monosymmetrical with respect to this axis. The flower resupinates at anthesis so as to present a single adaxial petal (Eichler 1878) (Figure 5.3). The three posterior petals (dorsal) therefore differ in having a smaller size, as well as having larger floral guides.

Wiegel and Lunau (2023) suggest that the high inter- and intra-individual variations in flower symmetry in the genus *Saxifraga* could indicate that this genus is at an intermediate state of evolution between polysymmetry and monosymmetry. They show that environmental factors such as gravity, as well as intrinsic factors, have an influence on flower symmetry. For instance, in the inflorescence, flowers oriented horizontally establish actinomorphy, whereas flowers oriented diagonally show monosymmetry.

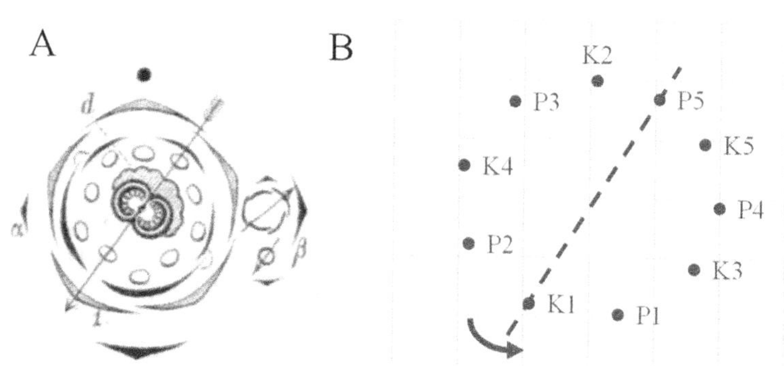

Figure 5.3. *(A) Floral diagram (Eichler 1878) of Saxifraga stolonifera syn. S. sarmentosa. (B) Model. Bracteoles are opposite. Arrow shows resupination*

The zygomorphy of *S. stolonifera* is positional because it arises late in ontogeny.

Organ	Angle (°)	Distance from the center
Adaxial source	90	3
Bract	270	3
β	0	2
α	180	2.5

Table 5.2. *Modeling parameters for Saxifraga stolonifera*

5.3. Malpighiales

With around 16,000 species, the order Malpighiales is one of the largest orders of angiosperms. Among Malpighiales, the family Malpighiaceae is sister to clades that possess actinomorphic flowers (Elatinaceae and Centroplacaceae) and has evolved from actinomorphic-flowered ancestors (Zhang et al. 2010).

Zhang et al. (2010, 2013) performed RT-PCR using locus specific primers to examine the expression of *CYC2* genes in Malpighiales.

5.3.1. *Elatinaceae*

Actinomorphic *Bergia* (Elatinaceae) species are found in temperate and tropical Eurasia and Africa; three species (two tropical and one tropical to temperate) occur in the Americas (Davis and Chase 2004).

In *Bergia texana* (Elatinaceae), the six *CYC2*-like genes exhibit no early expression but are expressed throughout the calyx and the corolla during later stages of development (Zhang et al. 2010) (Figure 5.4(B)).

The calyx of *Bergia* is quincuncial, with opposite bracteoles (Table 5.3). The correspondence between *CYC*-like gene expression and corolla symmetry (actinomorphy) respects Curie's symmetry principle, which is based on the symmetry of *CYC*-like gene expression.

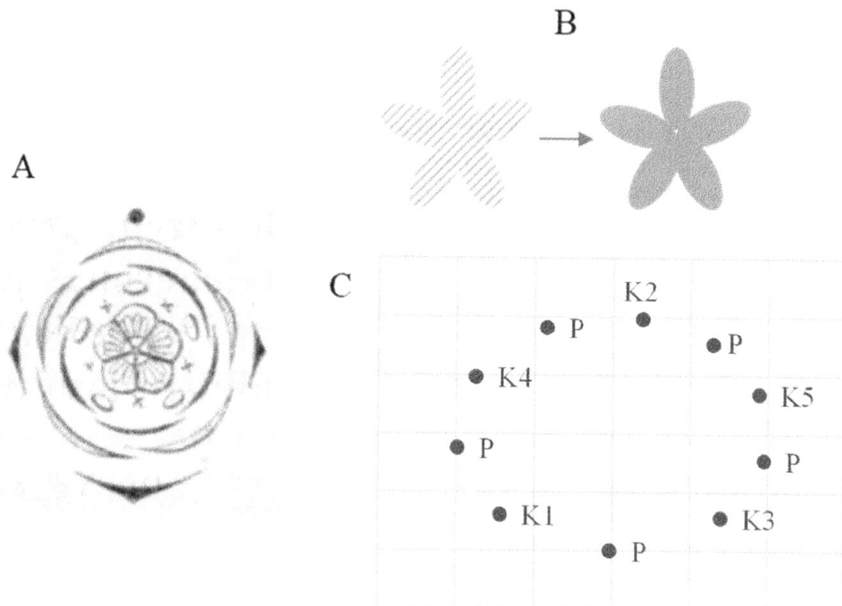

Figure 5.4. *(A) Floral diagram of Bergia ammannioides (Eichler 1875): calyx aestivation is quincuncial and the corolla is imbricate. (B) CYC2-like gene expression in the corolla of Bergia texana (Zhang et al. 2010). (C) Model of the arrangement of perianth organs in B. ammannioides. The petals are not numbered because their initiation is simultaneous*

Organ	Angle (°)	Distance from the center
Adaxial source	90	3
Bract	270	3
b	0	2
a	180	3

Table 5.3. *Modeling parameters for Bergia ammannioides*

5.3.2. *Malpighiaceae*

Malpighiaceae represent an excellent example of rapid radiation, during which there have been seven dispersal events from the New World to the Old World that have given rise to Old World Malpighiaceae. This tropical plant family exhibits a strong association between floral zygomorphy and insect pollinator attraction (Zhang et al. 2010).

The floral morphology of the more than 1,000 New World species of Malpighiaceae is very distinctive and highly conserved. The single dorsal banner petal is strongly differentiated from the other petals and appears to help orient and attract an extremely limited suite of pollinators, principally female bees of the tribes Centridini and Tapinotaspidini. Furthermore, the very narrow base of the petals provides bees access to oil glands, which are borne in pairs on the underside surface of the sepals (Zhang et al. 2010) (Figures 5.5(A) and (B)). The female bee clings to the (adaxial) banner petal at a preformed constriction and scratches the oil glands with its fore and mid legs: the lateral stigma–stamen complex touches its shoulder (Vogel 1990).

Based on phylogenetic analyses combined with fossil evidence and molecular divergence-time estimates, Davis et al. (2002) proposed that Malpighiaceae originated in northern South America, and that members of several clades repeatedly migrated into North America and subsequently moved via land connections across the North Atlantic into the Old World during episodes starting in the Eocene, when climates supported tropical forests.

The Old World lacks the oil–bee pollinators that visit New World Malpighiaceae, and these geographic transitions have resulted in alterations in both floral symmetry and petal morphology, as well as the loss of oil gland morphology, among Old World members of the family. Thus, the majority of Old World species possess flowers that are either truly actinomorphic or zygomorphic in a manner that is very different from the pattern exhibited by New World species. These zygomorphic Old-World species possess two dorsal petals instead of one (Figure 5.6(A)), two lateral petals and one ventral petal instead of two (Zhang et al. 2010).

New World Malpighiaceae

Stigmaphyllon and *Malpighia* are representative of native New World Malpighiaceae. Their calyx is quincuncial and their corolla ascending cochleate (Figure 5.5(A)). The flower is zygomorphic, the axis of symmetry is oblique at an angle of 36° with respect to the dorso-ventral axis during development and presents one adaxial petal (P5) at anthesis following a rotation, which also places K3 in the abaxial position (Eichler 1878) (Figure 5.5(D)).

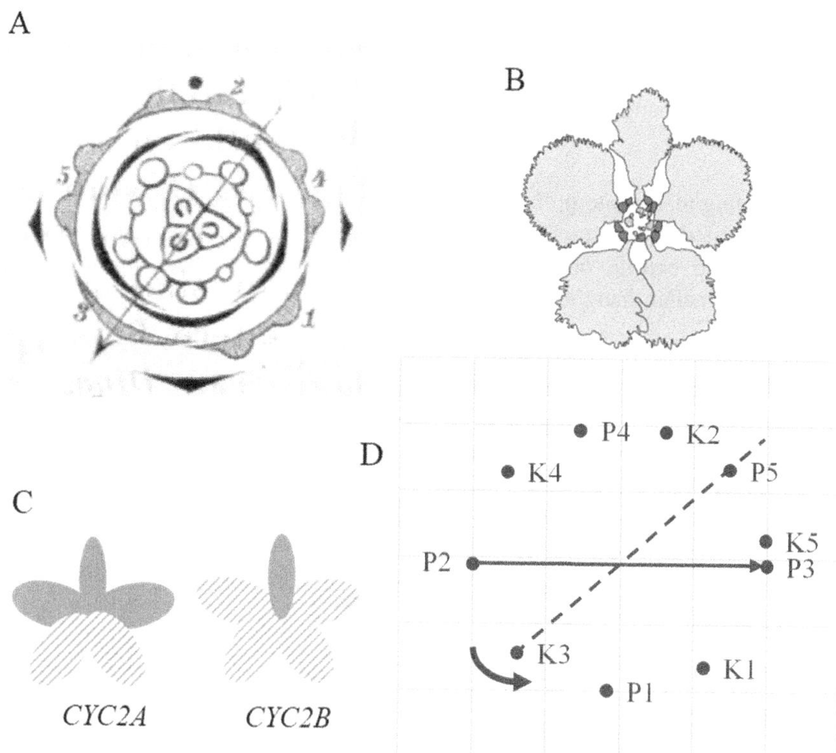

Figure 5.5. *(A) Floral diagram before rotation of Stigmaphyllon sp. (Eichler 1878). The bumps ("Drüsen") on the sepals are oil glands. (B) Mature flower of Stigmaphyllon redrawn from Papadopulos et al. (2013). (C) CYC2A and CY2B expression in the corolla of Byrsonima crassifolia (New World Malpighiaceae) (Zhang et al. 2010). Blue = expression of CYC-like genes in the floral meristem; striped = lack of expression. (D) Model of the arrangement of perianth organs in Stigmaphyllon. Parameters are listed in Table 5.4*

Organ	Angle (°)	Distance from the center
Adaxial source	90	2.8
Bract	270	3
a	10	2.5
b	180	1.8

Table 5.4. *Modeling parameters for Stigmaphyllon*

CYC2A-like and *CYC2B*-like genes are differentially expressed along the dorsoventral axis in the perianth of *Byrsonima crassifolia* (tropical America) and *Janusia guaranitica* (Bolivia, Brazil, North Argentina), "which mirrors their floral morphology" (Zhang et al. 2010) (expressing conformity to Curie's principle) (Figure 5.5(C)).

According to our interpretation, the initiation of the perianth is symmetrical with respect to the dorsoventral axis, with twinning of P2 and P3 and the New World Malpighiaceae corolla oblique zygomorphy is expressed later in development: P5 remains smaller than the other four petals.

5.3.3. *Old World Malpighiaceae*

Most of the Old World Malpighiaceae species have lost the distinctive banner-petal morphology, resulting in slightly zygomorphic (*Tristellateia*) or radially symmetrical (*Sphedamnocarpus*) flowers. Furthermore, the oil glands have either been completely lost, or modified to produce sugar-based rewards (Lobreau-Callen 1989). Old World Malpighiaceae primarily provide pollen and, less commonly, nectar to their pollinators (Vogel 1990).

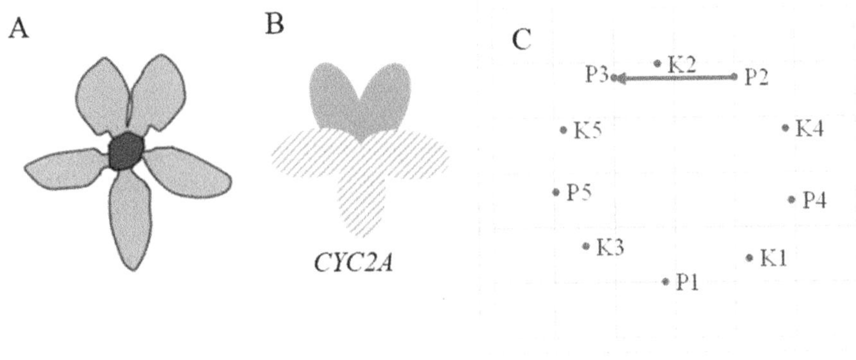

Figure 5.6. *(A) Mature flower of Tristellateia australasiae (Old World Malpighiaceae) redrawn from Zhang et al. (2010). (B) CYC2A expression in the corolla of T. australasiae (Zhang et al. 2010). (C) Model*

Tristellateia australasiae is found in eastern Africa, southeastern Asia and Madagascar. It has two dorsal, two lateral and one ventral petal (Figure 5.6(A)). *T. australasiae* has lost *CYC2B*; *CYC2A* is expressed during early development in the dorsal and two lateral sepals, and in the two dorsal petals (Figure 5.6(B)) (Zhang et al. 2010). This may contribute to the "shift in flower morphology" and reflect adaptations to a different pollination strategy. *Tristellateia* is visited by xylocopine bees whose reward for visiting the flowers appears to be pollen rather than oil (Zhang et al. 2010).

To model the arrangement of perianth organs in *T. austalasiae*, we forced the symmetry between P2 and P3 to take into account the expression of *CYC2A*-like gene (Figure 5.6(C); the parameters are listed in Table 5.5).

Organ	Angle (°)	Distance from the center
Adaxial source	90	3
Bract	270	3
a	10	2.8
b	180	2

Table 5.5. *Modeling parameters for Tristellateia australasiae*

Zhang et al. (2013) found two general *CYC2*-dependent mechanisms that may be responsible for the transition to actinomorphy from a zygomorphic ancestor in Old World Malpighiaceae: either the expansion of the expression domain of *CYC2* genes to the ventral region of the corolla (e.g. *Microsteria*; see Figure 5.7(A)) or the loss or significant reduction of *CYC2* expression (e.g. *Sphedamnocarpus*; see Figure 5.7(B)).

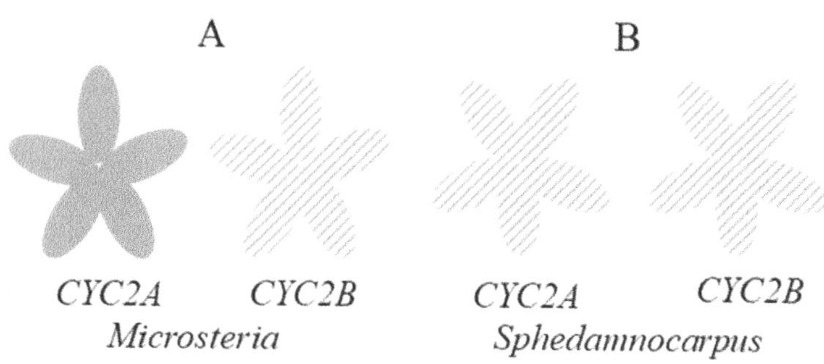

Figure 5.7. *CYC2A and CYC2B expression in the corolla of (A) Microsteria sp. and (B) Sphedamnocarpus (Zhang et al. 2013)*

The correspondence between *CYC*-like gene expression and corolla symmetry respects Curie's symmetry principle, which is based on the symmetry of *CYC*-like gene expression and not on the existence of this expression.

5.4. Ericales

5.4.1. *Balsaminaceae*

Impatiens

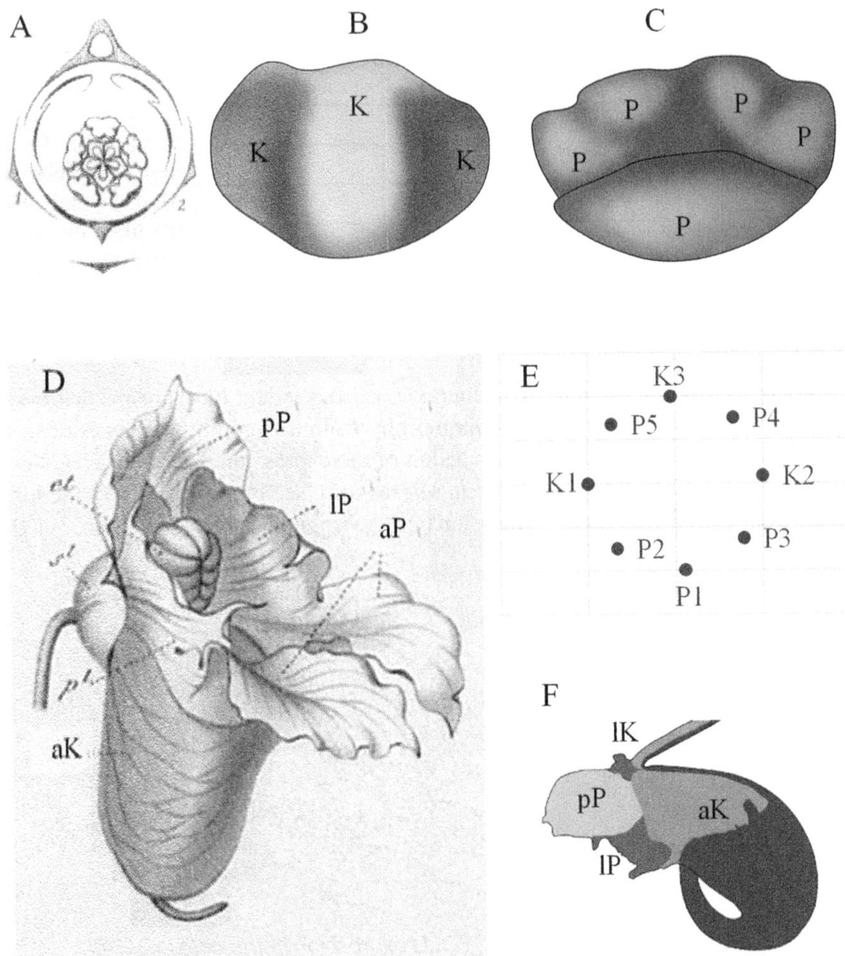

Figure 5.8. *(A) Floral diagram of Impatiens tricornis before anthesis (Eichler 1878). (B and C) Development of Impatiens niamniamensis (from Caris et al. 2006). (D) Drawing of Impatiens glandulifera (syn. royleana) after anthesis (Payer 1857). (E) Model of the arrangement of perianth organs in Impatiens at initiation. (F) Mature flower of Impatiens niamniamensis (from Caris et al. 2006)*

The flower of *Impatiens* (Balsaminaceae) is zygomorphic with three, occasionally five (where one pair is strongly reduced), sepals. The adaxial sepal is much larger, often colored and is characterized by a nectar spur arising late in development. The corolla consists of five petals: a large and often partly sepaloid (i.e. green) petal initiated on the abaxial side, and four lateral petals (Figure 5.8(A)). Flowers are resupinate at anthesis: thus, the spurred (adaxial) sepal is anterior (aK), the abaxial petal becomes posterior (pP), the other organs being lateral (lK and lP) (Figure 5.8(F)) (Caris et al. 2006).

Organ	Angle (°)	Distance from the center
Adaxial source	90	2.5
Bract	270	2.3

Table 5.6. *Modeling parameters for Impatiens*

The inflorescence is a compound raceme (see Figure 2(A) in Caris et al. 2006). The floral diagram from Eichler, as well as the SEM from Caris et al. (2006) (Figures 5.8(A) and (C)), show perfect bilateral symmetry early on from sepal initiation. Later, the spur becomes enormous, and the flower falls over at anthesis under the effect of its weight.

5.5. Geraniales

5.5.1. *Geraniaceae*

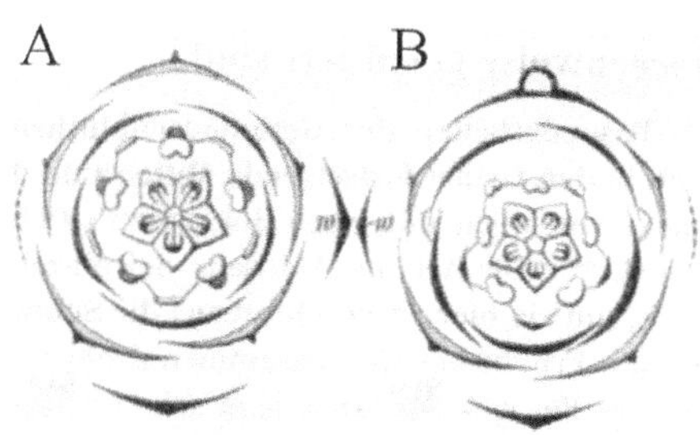

Figure 5.9. *Floral diagrams of (A) Erodium cicutarium and (B) Pelargonium zonale (Eichler 1878)*

In Geraniaceae, while flowers of *Geranium* and *Monsonia* are strictly polysymmetric, those of *Erodium* are actinomorphic. *Pelargonium* has strongly monosymmetric flowers with a corolla cochleate descending aestivation. The two dorsal petals differ from the ventral petals in shape, color and size (Figure 5.11(B)) (Ronse de Craene 2022).

Erodium

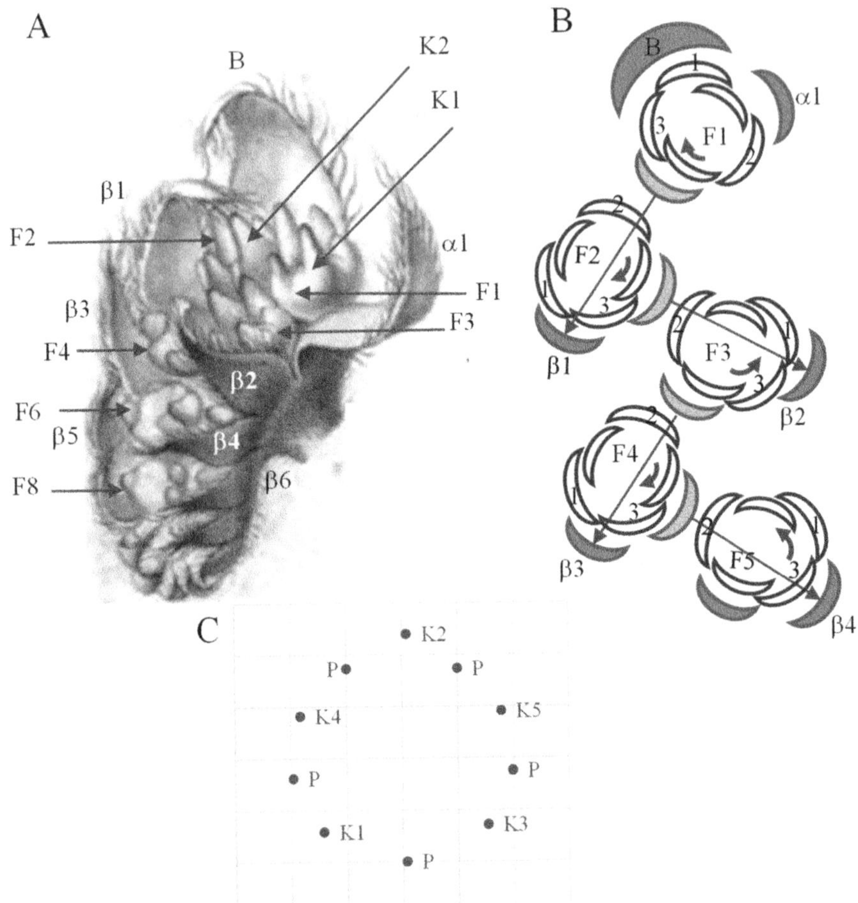

Figure 5.10. *(A) Scorpioid cyme of Erodium ruthenicum (syn. serotinum) (Payer 1857). (B) Modeling the cincinnus of Erodium ruthenicum (see Walch and Blaise 2023, Chapter 7). (C) Model of the arrangement of perianth organs in Erodium*

The raceme meristem of *Erodium* produces two bracts (B and B' (not shown)) and the terminal flowers F1 and F'1 (not shown) of two scorpioid cymes.

The sympodial meristem (SM) of F1 produces two bracteoles: α1 (which is sterile) and β1, in the axil of which an SM will produce a flower, F2, with only one bracteole, β2, and so on, thus forming a monobractal cincinnus. Bracteole position alternates between the right and the left of the flower (front view), resulting in alternate antidromous calices. Each cincinnus bears six to eight flowers, which are enveloped inside the fused bracteoles of the individual flowers (Figures 5.10(A) and (B)).

Eichler assigned one physical bracteole and one cryptic bracteole to each flower (Figure 5.9(A)). In our opinion, the latter represents the contact pressure with a neighboring floral meristem (FM).

Organ	Angle (°)	Distance from the center
Adaxial source	90	3
Bract	270	3
β	10	2
Cryptic	180	3

Table 5.7. *Modeling parameters for Erodium according to Eichler diagram (Figure 5.9(A))*

Corolla symmetry of *Erodium* is actinomorphic with quincuncial sepal aestivation (Figure 5.10(C)).

Pelargonium

The calyx of *Pelargonium* is quincuncial, like that of *Erodium*. The corolla becomes descending cochlear, an aestivation that is generally characteristic of strongly zygomorphic flowers. With the parameters used, petals did not need to be twinned: zygomorphy could be positional that corresponds to the formation of its spur (Figure 5.9(B)).

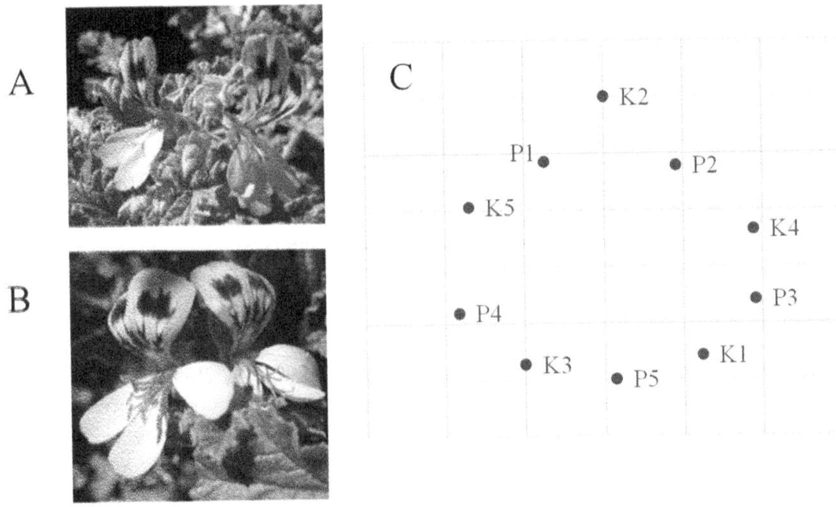

Figure 5.11. *(A and B) Dichasia of Pelargonium zonale showing zygomorphic flowers (photos by Solange Blaise). (C) Model of the arrangement of perianth organs in Pelargonium*

Organ	Angle (°)	Distance from the center
Adaxial source	90	3.4
Bract	270	3.2
β	190	1.5
Cryptic	20	2.3

Table 5.8. *Modeling parameters for Pelargonium according to Eichler diagram (Figure 5.9(B))*

5.6. Brassicales

5.6.1. *Brassicaceae*

Brassicaceae comprise around 350 genera whose members bear flowers with mainly polysymmetric corollas. However, in three genera, *Iberis*, *Teesdalia* and *Calepina*, the corolla is monosymmetric with differentially sized petal pairs. Inflorescences are generally racemose, and bracts and bracteoles are generally lacking (see Eichler "Unterdrückt") (Figure 5.12(C)).

Arabidopsis

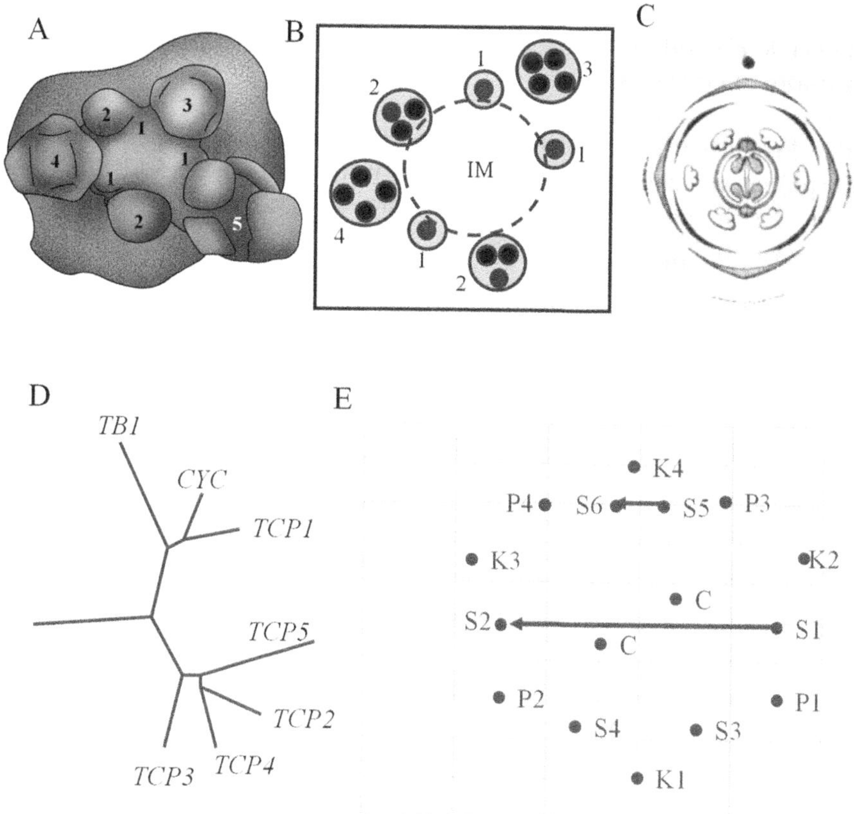

Figure 5.12. *(A) SEM of the main inflorescence apex of Arabidopsis (Smyth et al. 1990). (B) Model summarizing three stages during the development of the Arabidopsis floral meristem (FM): ephemeral bract (in red) initiation establishes abaxial and adaxial polarity (Table 5.9); sepals (in black) (Chandler and Werr 2014). Numbers correspond to the development stages of FM. (C) Floral diagram of actinomorphic Brassicaceae (Eichler 1878). (D) Tree showing the relationship between selected class II TCP proteins: TB1 in maize; CYC in Antirrhinum; TCPs in Arabidopsis (from Cubas et al. 1999). (E) Model of the arrangement of perianth organs, stamens and carpels in Arabidopsis*

The inflorescence of Arabidopsis *thaliana* (Brassicaceae) is a raceme: buds arise **acropetally** on the inflorescence meristem, each new FM arising around the apex at a 130°–150° angle to its preceding FM. The floral developmental sequence was divided into 12 stages by Smyth et al. (1990). Stage 1 begins with the initiation of a floral buttress on the flank of the apical meristem. Stage 2 starts when the flower primordium becomes separate from the meristem. Sepal primordia then arise (Stage 3) and grow to overlie the FM (Stage 4). Next, four petal and six stamen primordia appear (Stage 5) and are soon enclosed by the sepals (Figures 5.12(A) and (C)).

Sepals initiate unidirectionally in the following order: abaxial, the two lateral simultaneously and adaxial (Chandler and Werr 2014) (Figure 5.12(B)). According to some interpretations, an ephemeral bract emerges at the IM periphery during floral Stage 1 and is transformed into a sepal (Kwiatkowska 2006; Chandler and Werr 2014).

Organ	Angle (°)	Distance to the center (R) or plastochron ratio	Organ	Plastochron ratio
Adaxial source	90	R = 2	P4	1.1
Abaxial source	270	R = 5	S1	1
K1		1	S2	1
K2		1	S3	1
K3		1	S4	1
K4		1.1	S5	1
P1		1	S6	3
P2		1	C	1
P3		1	C	1

Table 5.9. *Modeling parameters for Arabidopsis*

CYC is a member of the *TCP* family of DNA-binding proteins, 24 of which have been identified in the genome of *Arabidopsis*. Of these, *TCP1* is the most similar to *CYC* and represents a likely ortholog (Figure 5.12(D)).

The model that fits the floral diagram exactly requires only the creation of twin stamens.

Iberis

Unlike *Arabidopsis thaliana* (Brassicaceae), which has four petals of equal size, *Iberis amara* (also Brassicaceae) produces two small dorsal and two large ventral petals (Busch and Zachgo 2007).

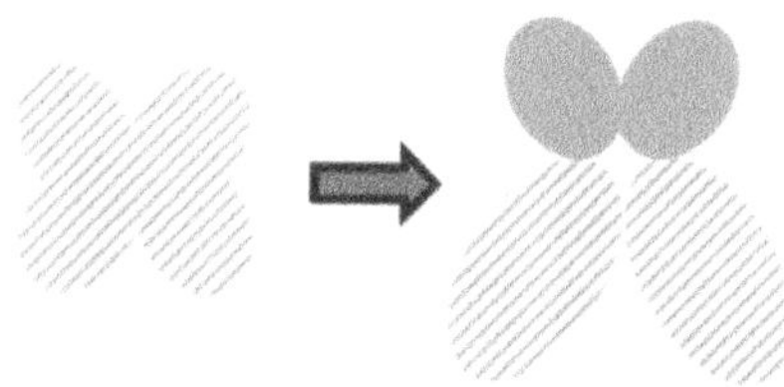

Figure 5.13. *Expression pattern of IaTCP1 in Iberis amara corolla showing positional zygomorphy (Busch and Zachgo 2007). The shape of the flower is stylized to represent the symmetry of the corolla at anthesis*

In *I. amara*, dorsal and ventral petal primordia initiate simultaneously and are of equal size. The difference in size between dorsal and ventral petals then increases gradually (from 1.6- to 3.7-fold) until anthesis (Busch and Zachgo 2007). Corolla monosymmetry in *Iberis* allows the flowers to sit closely together without covering neighboring flowers with protruding adaxial petals. Along with retarded early internode elongation and pedicel growth, this contributes to shaping the peculiar architecture of young *I. amara* inflorescences. The inflorescence has a corymboid structure with a flattened top and superficially resembles a single flower, an effect typically generated by the umbel of Apiaceae and the capitulum of Asteraceae (Busch and Zachgo 2007).

In situ hybridization experiments were carried out by Busch and Zachgo (2007) to determine the tissue-specific expression of *IaTCP1* (the homolog of *Arabidopsis TCP1*) in *I. amara*. No asymmetric expression could be detected in young FMs. After the beginning of stamen differentiation, low *IaTCP1* expression was observed in young petals and stamens. However, as floral organ differentiation advanced further, a stronger expression was found in the adaxial petals compared to the abaxial petals (Figure 5.13) (Busch and Zachgo 2007). Thus, the zygomorphy of *Iberis* is positional.

5.6.2. *Tropaeolaceae*

The monogeneric family Tropaeolaceae, comprising circa 90 species native to the Americas, is the only family of the highly diversified Brassicales with a floral medial nectar spur (Ronse De Craene and Smets 2001).

Tropaeolum

The floral ontogeny of several species of *Tropaeolum* was studied by Ronse de Craene and Smets (2001) and Martinez-Salazar et al. (2021) using a scanning electron microscopy. Floral symmetry changes during floral development from "radial" symmetry at petal initiation to strong bilateral symmetry at anthesis due to the formation of an adaxial spur.

Figure 5.14. *(A) Floral bud of Tropaeolum majus showing the spur from Frank Vincentz (2007). File licensed under the Creative Commons Attribution license. (B) Mature flower of T. majus from Jose Luis Cernadas Iglesias (2009). File licensed under the Creative Commons Attribution license. (C) Floral diagram of T. majus (descending cochleate aestivation) (Eichler 1878), including cryptic bracteoles*

Early floral development

Flowers are arranged in compressed racemes showing (3, 5) Fibonacci spirals (Figure 5.15(A)). No bracteoles are formed: Eichler imagined cryptic bracteoles (Figure 5.14(C)), however we propose that neighboring bracts exert contact pressure on developing flowers (Figure 5.15(A), e.g., pressure by B5 (+8), B8 (+5) and B10 (+3) on flower meristem FM13 during sepal initiation).

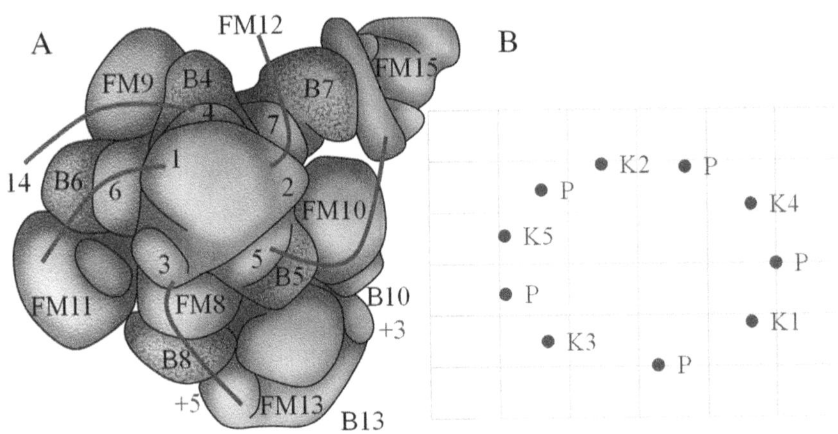

Figure 5.15. *(A) Apical view of the inflorescence meristem of Tropaeolum peregrinum with the spiral sequence of initiation of flower primordia (4–7 and FMn) in the axil of bracts (1–3 and Bn); +3 and +5 = parastichies. Redrawn from Ronse de Craene and Smets (2001). (B) Modeling the initiation of Tropaeolum peregrinum perianth organs*

The five sepal primordia are initiated in a quincuncial arrangement. Sepals grow and touch each other without being fused, then petal primordia arise simultaneously (Ronse de Craene and Smets 2001).

Organ	Angle (°)	Distance from the center
B5	90	3
B13	270	2.5
B10	20	1.9
B8	160	2.5

Table 5.10. *Modeling parameters for Tropaeolum peregrinum for FM13*

Later stages of development

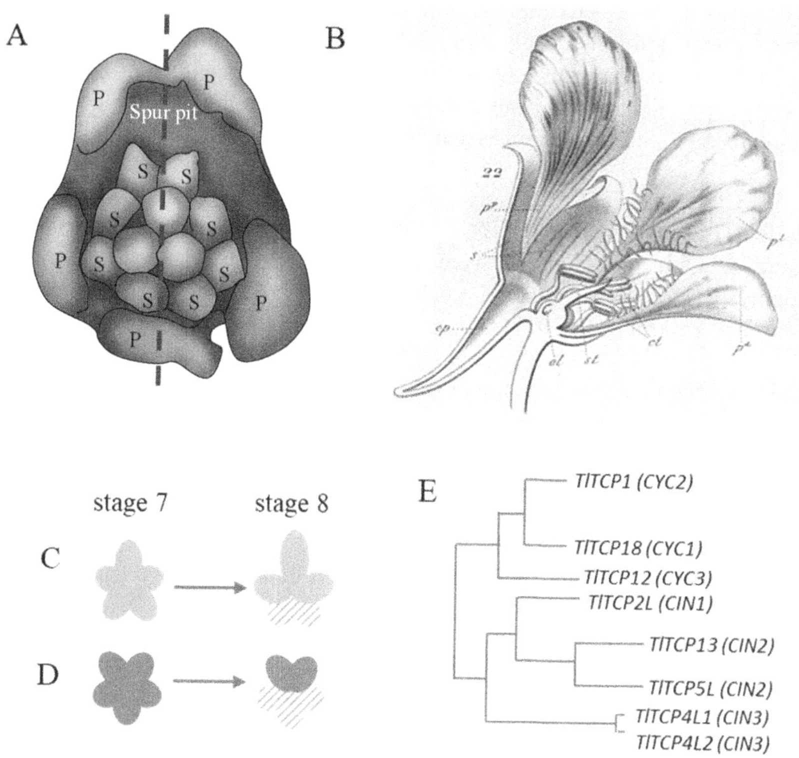

Figure 5.16. *(A) Apical view of a Tropaeolum majus flower with the spur forming as a small pit (Ronse de Craene and Smets 2001). (B) T. majus at anthesis showing the spur (Payer 1857). (C and D) Expression of TCP1 (CYC2) gene in Tropaeolum longifolium in (C) sepals and (D) petals. (E) Phylogenetic tree of TCP genes in T. longifolium from Martínez-Salazar et al. (2021)*

At initiation, petals are of equal size, but during spur formation a gradual differentiation in size and morphology occurs when the dorsal petals become larger than the ventro-lateral petals. The result is a shift from actinomorphy to strong dorsoventral monosymmetry. Stamens are arranged symmetrically on either side of the dorsoventral axis (Figures 5.14(C) and 5.16(A)) (Ronse de Craene and Smets 2001). The rapid growth of the spur displaces the free lobes of both the dorsal sepal and petals away from the floral axis (Martínez-Salazar et al. 2021).

Martínez-Salazar et al. (2021) reconstructed the evolution of *TCP* genes in *Tropaeolum* and assessed their expression in *Tropaeolum longifolium* to determine their contribution to the differential adaxial–abaxial growth of the floral tube during spur formation (Figures 5.16(C) and (D)). They studied the spatiotemporal gene expression patterns at floral developmental stages that coincided with the developmental events reported in *T. majus* and *T. peregrinum* by Ronse De Craene and Smets (2001). Martínez-Salazar et al. (2021) defined the stages of flower development as follows: Stages 1 and 2: sepal and petal primordia initiation; Stages 3 and 4: stamen primordia and gynoecium initiation; Stage 5: spur initiation; Stages 6 and 7: spur elongation (1 cm); Stage 8: spur is 2 cm long and its coloration shifts from green to pink before anthesis.

TlTCP12 and *TlTCP2L* were only expressed in the ventral sepals at Stage 7, while *TlTCP1* (*CYC2*) and *TlTCP12* were only expressed in the dorsal sepals at Stage 8. Only three genes were exclusively expressed in the dorsal petals, namely, *TlTCP4L2* at Stage 7 and *TlTCP1* (*CYC2*) and *TlTCP12* at Stage 8 (Martínez-Salazar et al. 2021) (Figures 5.16(C) and (D)).

Modeling did not require the twinning of petals (Figure 5.15(B)). Zygomorphy is positional, which corresponds to the expression of *TCP* genes: the dorsoventral asymmetry of *TlTCP1* expression only appears at Stage 8 with the elongation of the spur. Because of their growth, petal aestivation changes from valvate to descending cochleate. At anthesis, the weight of the spur counterbalances that of the rest of the flower (Figure 5.14(A)).

5.7. Solanales

5.7.1. *Solanaceae*

Most Solanaceae species have an actinomorphic corolla, but in some species the limbs can be strongly zygomorphic with deeply fringed petals (e.g. *Schizanthus*) or have distinct color patterns in the dorsal region (e.g. *Salpiglossis*) or be slightly zygomorphic with petal size varying of along the dorsoventral axis (e.g. *Nicotiana*) (Zhang et al. 2017).

The early floral orientation in zygomorphic Solanaceae has been the subject of debate. The two fused carpels were described by Eichler (1875) as oblique during development, being 36° off the dorsoventral axis defined by a line that passes through the inflorescence axis and the flower-subtending bract. At anthesis the position of the single dorsal petal (like in Caesalpinieae) was considered by Eichler to be the result of a 36° resupination, while this arrangement is constitutional in Caesalpinieae (Figure 5.17(A)).

Dichasial cymes are relatively rare in Solanaceae but can be found for instance in *Datura*. Solanaceae possess predominantly **scorpioid cymes** (also called **cincinni**). This cymose branching pattern has evolved independently in other families such as Boraginaceae, Caryophyllaceae and Gentianaceae. Two main forms of inflorescence exist in Solanaceae: inflorescences where flowers are loosely arranged and inflorescences where flowers are tightly clustered. Loosely arranged inflorescences (e.g. *Datura* and *Petunia*) generally have two conspicuous bracteoles associated with each flower and long internodes separating these flowers. In this type of inflorescence, there are usually vegetative phyllomes between the flowers. In tightly clustered inflorescences (e.g. *Solanum* (without bracteoles), *Nicotiana* (one bracteole) and *Salpiglossis* (two bracteoles but only one is fertile)), the bracteoles and internodes are usually highly reduced, and the flowers are clustered together (Barboza et al. 2016; Zhang et al. 2022a).

The shift from loosely arranged inflorescences with extended flowering periods to tightly clustered inflorescences with synchronized flowering is thought to be an adaptation to a shortened growing season. For example, in clades with both types of inflorescences, such as Scrophulariaceae, Fabaceae, Campanulaceae and Caprifoliaceae-Valerianoideae, species with loosely arranged inflorescences usually occur in moist habitats and have an extended flowering season, while species with tightly clustered inflorescences are more likely to be found in either alpine or xeric habitats and have a short flowering season. Inflorescences with clustered flowers likely enable the plants to produce flowers and seeds in a shorter favorable period to avoid harsh environmental conditions such as in the cold or droughts (Zhang et al. 2022a).

Zhang et al. (2022a) challenged Eichler's floral diagram (Figure 5.17(A)) and proposed that the α bracteole has the same absolute angle (θ) as sepal K1. From the point of view of phyllotaxis, this position is impossible because it would assume zero inhibition generated by this bracteole (or infinite distance from the center of the flower). Furthermore, this is contradicted by the SEM of Souer et al. (1998) (Figure 5.17(B)) where sepal K1 is far from bracteole α3.

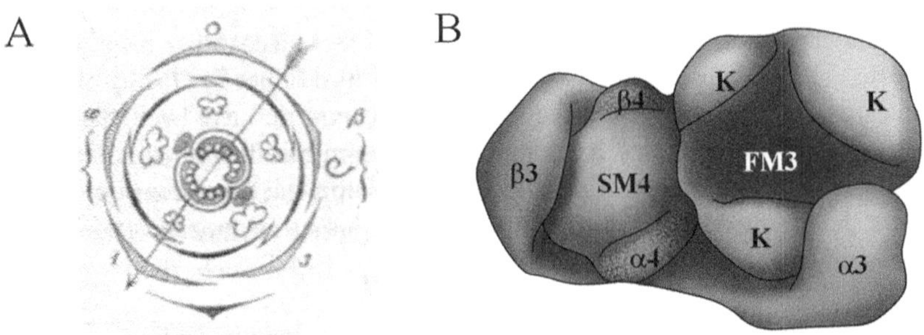

Figure 5.17. *(A) Floral diagram of Petunia nyctaginiflora before anthesis (Eichler 1875; the hook on the right represents the continuation of the cincinnus). (B) Petunia inflorescence showing floral meristem FM3 (on which the first three sepal primordia are visible) and the younger SM4 (redrawn from Souer et al. (1998))*

Both *Atropa beladonna* and *Petunia*, which have two bracteoles per flower (only β is fertile), have been previously studied by Walch and Blaise (2023, Chapter 7). Here, we studied *Solanum* that has clustered monochasial inflorescences without bracteoles and *Salpiglossis* with two bracteoles per flower (only β is fertile) and oblique zygomorphy.

Solanum

The tomato plant (Solanaceae) has a compound shoot formed from reiterated units, each composed of a few leaves and a scorpioid cyme without bracteoles. Each unit starts with a sympodial shoot meristem (SYM) that produces leaves before entering floral transition. A new SYM and a first flower (F1 on Figure 5.18(C)) are initiated. An SM arises and produces a new SM before differentiating into an FM. This pattern is reiterated as subsequent SMs develop perpendicularly to one another, producing the typical zigzag pattern of cincinnus (from Lippman et al. (2008)).

Calyx development in F3 shows that it is affected by the contact from flowers F1, F2, F4 and F5, whose pressure give the calyx its shape. At earlier F4 initiation stage, contact pressure is not complex enough to influence sepal location (Figure 5.18(C)).

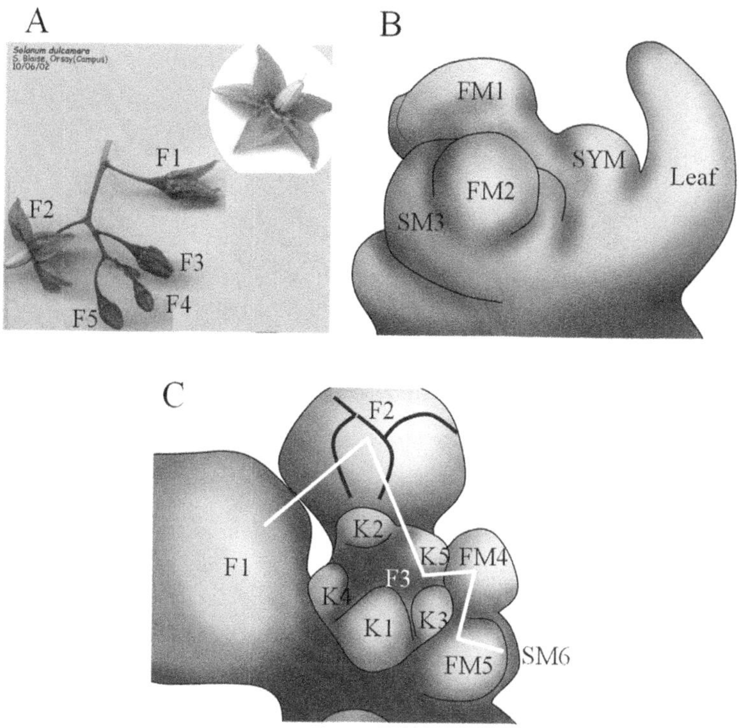

Figure 5.18. *(A) Solanum dulcamara from the Flore Virtuelle d'Orsay. (B and C) Cincinnus development in Solanum pimpinellifolium from Welty et al. (2007) showing zigzag initiation: SYM = sympodial shoot meristem of the next shoot unit; SM = sympodial meristem of the next flower; FM = floral meristem; F = flower*

Salpiglossis

According to Eichler (1875), *Salpiglossis* (Solanaceae) presents oblique zygomorphy (Figure 5.19(B)) along an axis inclined at 36° from the dorsoventral axis. The single dorsal petal evident in mature flowers (Figure 5.19(A)) would be the result of a 36° resupination at anthesis. During development, the gynoecium is also oblique and becomes symmetrical with respect to the vertical axis at anthesis.

According to Ampornpan and Armstrong (2002), at initiation, the calyx is quincuncial, the sepals K1 and K3 being located on either side of the bract. Petal initiation is valvate. During development, sepal K1 becomes epibractate, the rotation of the flower axis taking place before anthesis. Corolla aestivation is descending cochlear, and the adaxial position of petal P1 is not the result of resupination (rotation of the pedicel as in orchids and Lobeliaceae) since it happens before anthesis (Figure 5.19(D)).

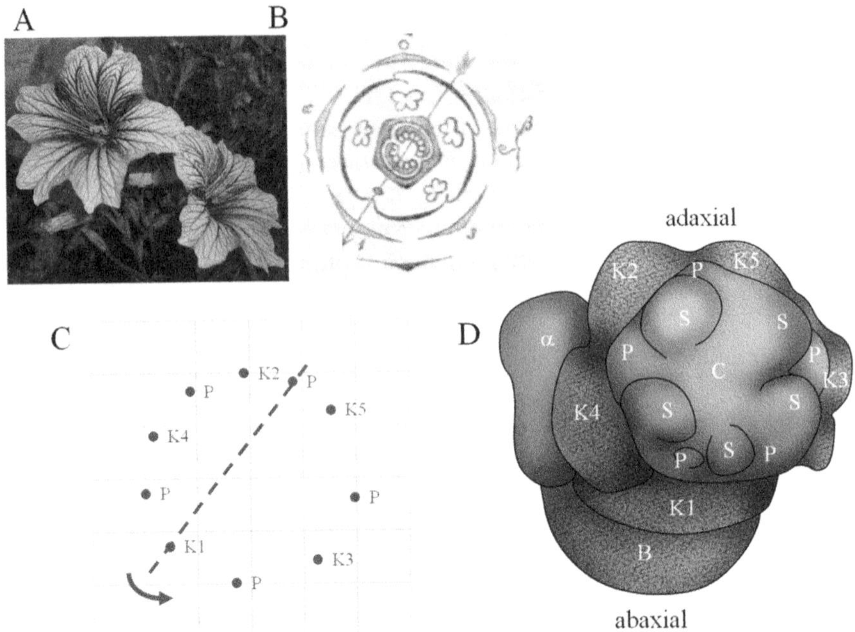

Figure 5.19. *(A) Mature flower of Salpiglossis sinuata from Salicyna (2015). File licensed under the Creative Commons Attribution license. (B) Floral diagram before anthesis of S. sinuata (Eichler 1875). The hook on the right represents the continuation of the cincinnus. (C) Modeling the initiation of perianth organs in S. sinuata. (D) SEM before anthesis of S. sinuata redrawn from Ampornpan and Armstrong (2002)*

Organ	Angle (°)	Distance from the center
Adaxial source	90	3
Bract	270	2.3
β	10	1.6
α	160	2.1

Table 5.11. *Modeling parameters for Salpiglossis sinuata*

The inflorescence of Salpiglossis is scorpioid with two bracteoles per flower, with development in β (Figure 5.19(B)). Modeling did not require the twinning of petals: zygomorphy is positional (Figure 5.19(C)).

5.8. Asterales

5.8.1. *Asteraceae*

The capitulum (pseudanthium or "false flower"), a dense cluster of flowers and bracts that has evolved to mimic a single flower, is a characteristic feature of the sunflower family Asteraceae. The evolution of the capitulum is thought to be a key innovation responsible for the evolutionary success of Asteraceae, making it one of the largest plant families.

A typical capitulum consists of many flowers (florets) and phyllaries (modified bracts) compressed into a single structure (e.g. *Matricaria inodora*, Figures 5.20(A) and (B)). Depending on their distribution in the clades, we distinguish radiates whose capitulum has rays on the periphery and discs inside (e.g. *Matricaria*), ligulates having only rays (e.g. *Taraxacum*), and discoids which only have discs (e.g. *Senecio arvensis*).

Ray florets have bilateral symmetry with fused ventral petals protruding like a tongue (ligule: Figure 5.22(B)), whereas disc florets have radial symmetry. In *Senecio,* discs have no bracts, whereas in *Anthemis* they do (Harris 1995). The formation of phyllaries and florets is asynchronous: it is most commonly acropetal (forming from the margin to center of the capitulum), but in some species it is bidirectional (Zoulias et al. 2019).

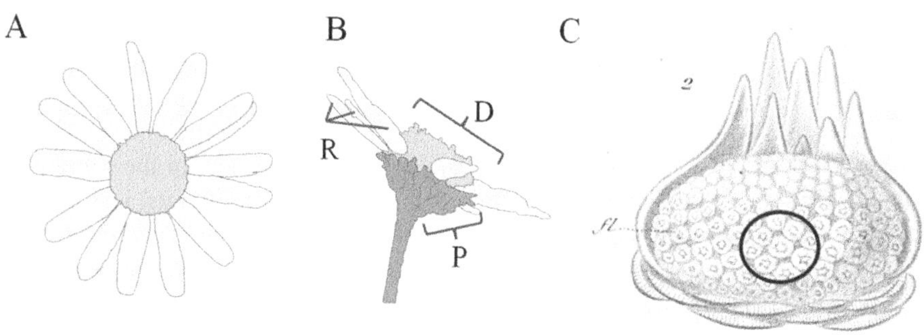

Figure 5.20. *Capitulum morphology and development in Matricaria inodora (A and B). P = phyllary or involucral bracts; R = ray floret; D = disc floret. Redrawn from Zoulias et al. (2019). (C) Capitulum of Calendula officinalis (Asteraceae) (Payer 1857). The circle shows the neighboring disc FMs of one of them located in the center of this circle*

Senecio

In Asteraceae (e.g. *Helianthus annuus*), the capitulum can bear hundreds of disc florets in Fibonacci spirals of up to (89, 144) mode. Our static model simulated the arrangement of these florets (as in Figure 5.20(C): see Walch and Blaise 2023, Chapter 3).

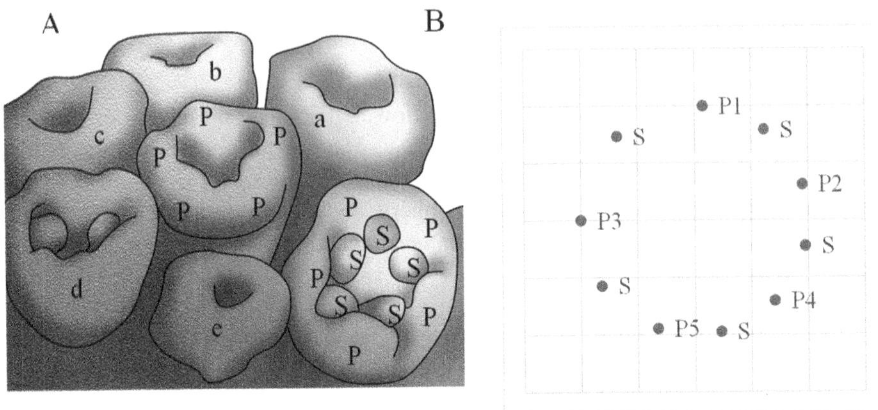

Figure 5.21. *(A) Development of disc florets of Packera glabella (Senecio glabellus) (butterweed) redrawn from Harris (1995). P = petals forming a corolla tube; S = stamen. (B) Model of the actinomorphic corolla and stamen arrangement in Packera*

We made the assumption that in the absence of bracteoles within racemes such as that of *Antirrhinum majus*, the forces shaping the perianth are not due to biochemical activity but instead due to contact pressure from the bracts subtending the flowers above each developing flower in the inflorescence. In the same way in Asteraceae, we assumed that the forces shaping the disc florets is the contact pressure between them (a–e in Table 5.12 and Figure 5.21).

Organ	Angle (°)	Distance from the center
a	45	2
b	90	3
c	135	2
d	225	2
e	270	3

Table 5.12. *Parameters for modeling the disc floret of Packera glabella (Senecio glabellus). (a–e) = neighboring flower location (Figure 5.21)*

Initiation of a regular pentamerous flower does not require that the position of neighboring flowers has specific symmetry (Figure 5.21; Table 5.12). The modeling of actinomorphic disc florets did not require the twinning of petals in accordance with the genetic analysis of Garcês et al. (2016) (Figure 5.22(A)).

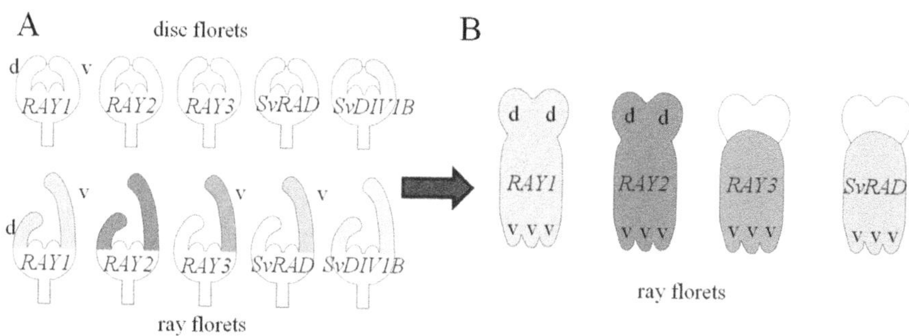

Figure 5.22. *Symmetry gene expression in developing disc and ray florets in Senecio. There is never an expression of RAY and RAD genes in disc florets. (A) At Stage 5, as petals are differentiated in ray florets, RAY3 and SvRAD are expressed in the ventral petals of ray florets and* ***promote ventral petal elongation****. (B) At Stage 8, RAY genes are expressed in zygomorphic ray florets. However, only a subset of RAY genes such as RAY3 is specifically expressed in ventral petals and promotes ventral petal growth. d = dorsal petal; v = ventral petal (redrawn from Garcês et al. 2016)*

Garcês et al. (2016) investigated the role of *CYC* and *RAD* in the development of ray and disc florets within a capitulum of Asteraceae by isolating homologs of these genes in *Senecio vulgaris*. The capitulum consists of bilateral (zygomorphic) ray florets and radial (actinomorphic) disc florets. Results showed that *RAY3* mRNA transcripts accumulated in the emerging ray floret primordia of the Stage 1 capitulum. Once disc and ray florets were formed (Stages 3–4), high and **uniform expression** was detected for all three genes exclusively in ray florets but absent in disc florets. In Stage 5 of capitulum development, *RAY3* and *SvRAD* were still highly expressed in ray florets but **only in the ventral petals** (Figure 5.22).

While floral symmetry genes appear to have a conserved role in petal growth in both *Senecio* and *Antirrhinum*, the expression domains are divergent, allowing ventral petal elongation in *Senecio* versus dorsal petal elongation in *Antirrhinum*. But in both cases, the expression of the *CYC*-like gene is asymmetric with respect to the transverse axis, which corresponds to zygomorphy according to Curie's principle. As petals are differentiated (Stage 5), *RAY3* accumulates in the ventral ray petals, which leads to rapid ventral petal elongation, thus to the bilateral symmetry of ray floret. The dorsoventral asymmetry of the ray petals is late, like the only ventral expression of the *RAY3* gene. Zygomorphy is positional.

5.9. Dipsacales

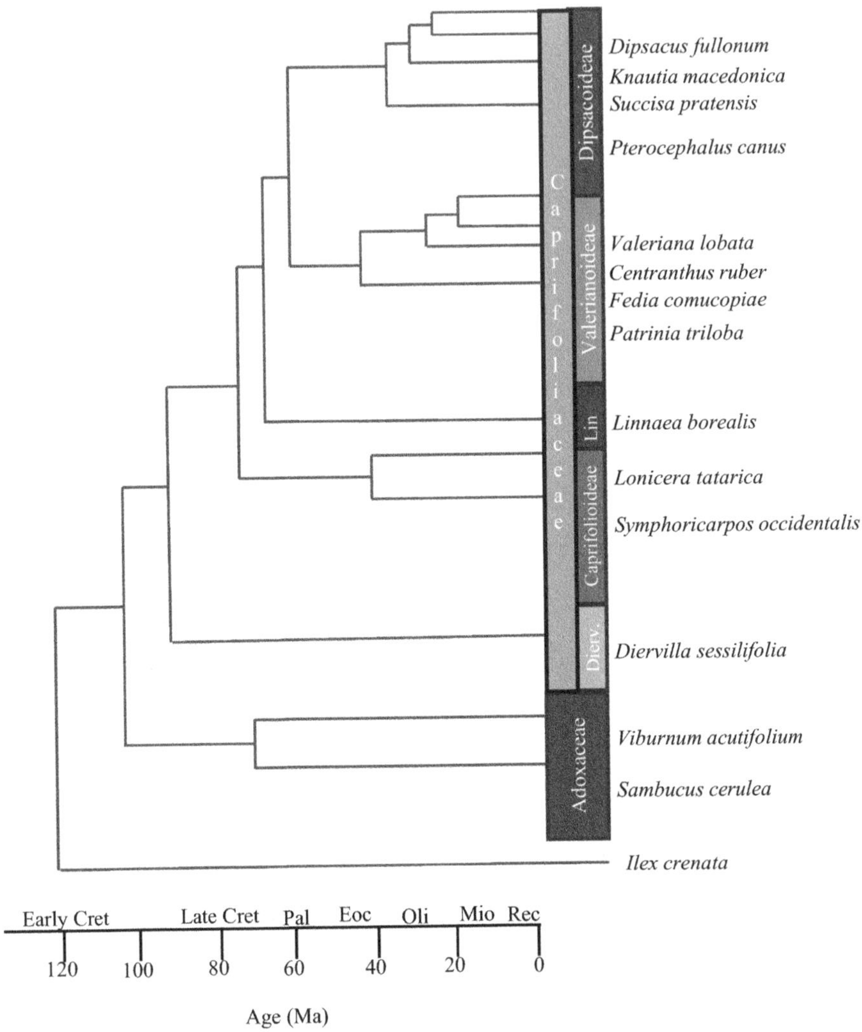

Figure 5.23. *Time-calibrated tree constructed using nuclear data of Dipsacales showing divergence time estimates (simplified from Lee et al. (2021)). Dier = Diervillioideae; Lin = Linnaeeae*

According to Lee et al. (2021), the first Dipsacales were woody plants with simple opposite leaves, and their flowers were sympetalous with five stamens alternating with five corolla lobes. **Adoxaceae** then probably evolved smaller rotate corollas and fleshy fruits with a reduced number of seeds enclosed in endocarps. **Caprifoliaceae**, in contrast, evolved long tubular bilaterally symmetrical corollas (with several reversals to radial symmetry), a long style with a capitate stigma, and nectaries in the form of small hairs at the base of the corolla tube. Caprifoliaceae is also marked by a reduction from five to four stamens. In this clade, there is also a reduction to one functional carpel (by abortion of two of the three carpels) and the production of an achene fruit. Within this clade, supernumerary inflorescence bracts were variously modified in relation to dispersal, and an epicalyx closely enveloping the ovary probably evolved independently several times.

The putative ancestor of Dipsacales was radially symmetrical (*Ilex*, Aquifoliaceae). The order Dipsacales is divided into two main clades: Adoxaceae and Caprifoliaceae.

5.9.1. *Adoxaceae*

Adoxaceae (including *Viburnum*, *Sambucus* and *Adoxa*) have radially symmetrical flowers with small calyx lobes and rotate corollas (Howarth et al. 2011). *Viburnum* petal aestivation is imbricate (Figure 5.24(B)).

Figure 5.24. *(A) Viburnum opulus from Flore virtuelle d'Orsay. (B) Floral diagram of Viburnum lantana (Eichler 1875). (C) Expression of DipsCYC2A and DipsCYC2B in Viburnum plicatum (Howarth et al. 2011)*

Two copies of genes from the *CYC2* clade, *DipsCYC2A* and *DipsCYC2B*, are found in *Viburnum plicatum* and are the product of a duplication event that was independent from the duplication of *CYC2* genes in Caprifoliaceae. Both copies were found to be expressed in whole buds of the central (fertile) flowers, sterile marginal flowers and leaves. Expression was also seen in all five petals (Figure 5.24(C)) (Howarth et al. 2011).

These correspondence between *CYC*-like gene expression and perianth symmetry respects Curie's symmetry principle for pentamerous flowers.

5.9.2. *Diervillioideae*

Caprifoliaceae (including Diervillioideae, Caprifolioideae, Linnaeoideae, Valerianoideae and Dipsacoideae) typically have larger calyx lobes and tubular corollas. The transition to bilateral symmetry in Caprifoliaceae involved the duplication of an ancestral *DipsCYC2* gene and the loss of expression of both of these copies in the ventral petal (Howarth et al. 2011).

Diervilla

One of the first diverging lineages in Caprifoliaceae, the genus *Diervilla* has weakly bilaterally symmetrical flowers with a 2:3 form. *DipsCYC2A* and *DipsCYC2B* are both strongly expressed in both dorsal and lateral petal lobes. Expression of either gene is extremely weak or absent in the ventral petal (Figure 5.25) (Howarth et al. 2011).

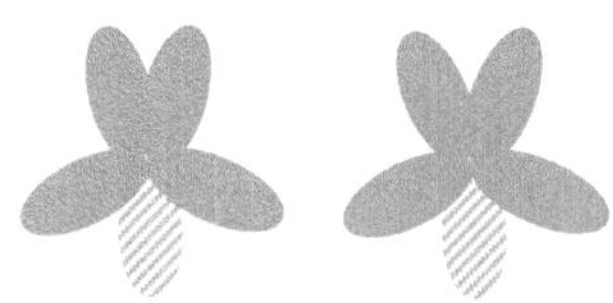

Figure 5.25. *Expression of DipsCYC2A and DipsCYC2B in Diervilla sessilifolia (Howarth et al. 2011)*

The correspondence between *CYC*-like gene expression and perianth symmetry respects Curie's symmetry principle for pentamerous flowers.

5.9.3. *Caprifolioideae*

Symphoricarpos

In *Symphoricarpos albus* (Caprifolioideae), sepal initiation is unidirectional from the adaxial to the abaxial side (Figure 5.26(B)). Within the emerging sepal whorl, a ring-like zone develops. The five petal primordia are initiated on this ring, which will form a corolla tube (Figure 5.26(A)) (Roels and Smets 1996). The flower is actinomorphic.

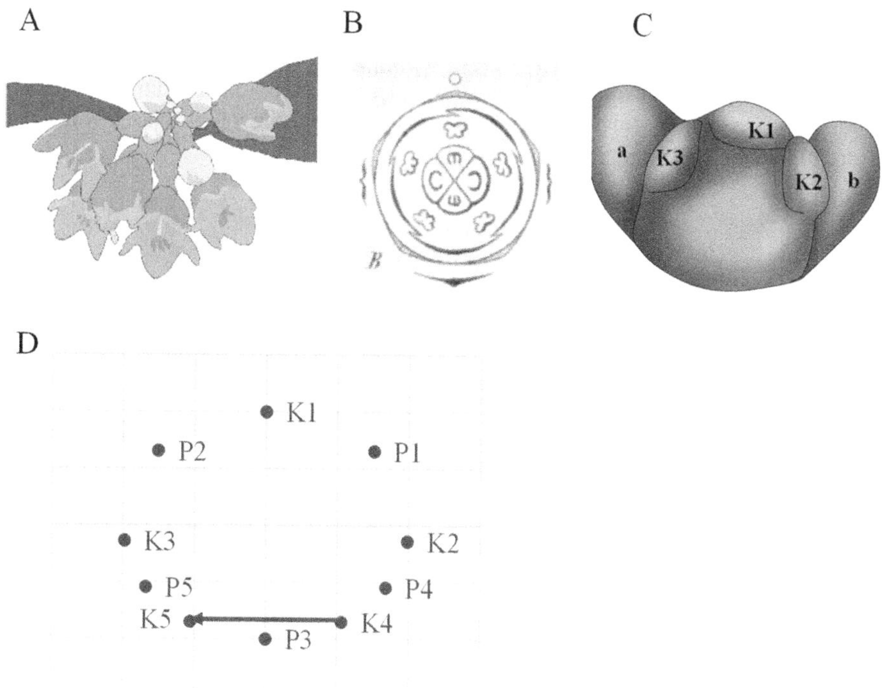

Figure 5.26. *(A) Mature flowers of Symphoricarpos albus showing corolla tubes. Redrawn from ArtMechanic (2005; file licensed under the Creative Commons Attribution license). (B) Floral diagram of S. racemosa (Eichler 1875). (C) Unidirectional sepal initiation in S. albus (redrawn from Roels and Smets (1996)). (D) Model of the arrangement of perianth organs in S. albus*

Organ	Angle (°)	Distance from the center
Adaxial source	90	4
Bract	270	1.5
a	340	2.5
b	200	2.5

Table 5.13. *Modeling parameters for Symphoricarpos albus*

The sources of inhibition are almost bilaterally symmetrical, however the symmetry of sepals K4 and K5 must be forced, otherwise sepal K5 would be located where P2 is located.

Lonicera

The inflorescence of *Lonicera tatarica* is a two-flowered dichasium where the central flower is aborted (Figure 5.27(A)). The five sepal primordia are initiated in rapid succession. A ring-like zone then forms within the sepal whorl where petals are initiated (Roels and Smets 1996).

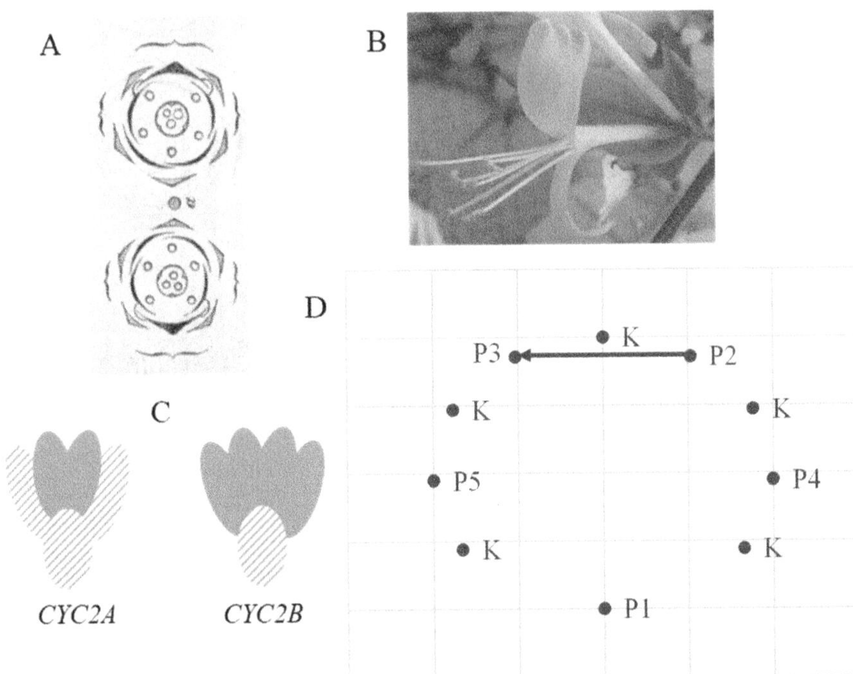

Figure 5.27. *(A) Inflorescence diagram of Lonicera tatarica (a dichasium) (Eichler 1875). (B) Flower of L. periclymenum showing 4:1 arrangement (photos by: Solange Blaise). (C) DipsCYC2A and DipsCYC2B expression in L. periclymenum (Howarth et al. 2011). (D) Model of the arrangement of perianth organs in L. tatarica*

Although all *Lonicera* species have a 4:1 flower form, they vary from being weakly bilaterally symmetrical (*L. sempervirens*) to very strongly zygomorphic, with the two dorsal and two lateral petals almost entirely fused and pointing upward (*L. periclymenum*) (Howarth et al. 2011).

In *L. periclymenum*, *DipsCYC2A* is expressed in the dorsal petals but not in the lateral or ventral petals. *DipsCYC2B* is expressed in the dorsal and lateral petals but not in the ventral petal. *L. periclymenum* is strongly bilaterally symmetrical with deep slits between the ventral petal and the upper four petals (Howarth et al. 2011) (Figure 5.27(B)). The correspondence between *CYC*-like gene expression and perianth symmetry respects Curie's symmetry principle for pentamerous flowers.

Organ	Angle (°)	Distance from the center
Adaxial source (a)	90	4
Bract	270	2
b	0	2.4
a	180	2.4

Table 5.14. *Modeling parameters for Lonicera tatarica*

The sources of inhibition are bilaterally symmetrical; however, the symmetry of petals P2 and P3 must be forced, otherwise petal P3 would be located where P5 is located. Modeling thus reproduces the multidirectional initiation of petals described by Eichler (1875) (Figure 5.27(A)).

5.9.4. *Valerianoideae*

Centranthus

The inflorescence of *Centranthus ruber* is a thyrse bearing scorpioid cymes (Figure 5.28(A)). The scorpioid cyme develops in β, α, sterile, being opposite it. The apex of the raceme generates opposite bracts in the axils of which the FMs are initiated. On a ring-like zone (lamina reduced to scales) of the floral apex, one stamen primordium is initiated in a nearly adaxial position. Then, five broad petal primordia are initiated on this ring-like zone (Figure 5.28(B)) (Roels and Smets 1996).

The flowers of *Centranthus* have a single stamen that is not in the median plane, whereas the conspicuous spur is in the median plane. Thus, the flowers appear asymmetrical at anthesis (Figure 5.28(A)) (Endress 1999).

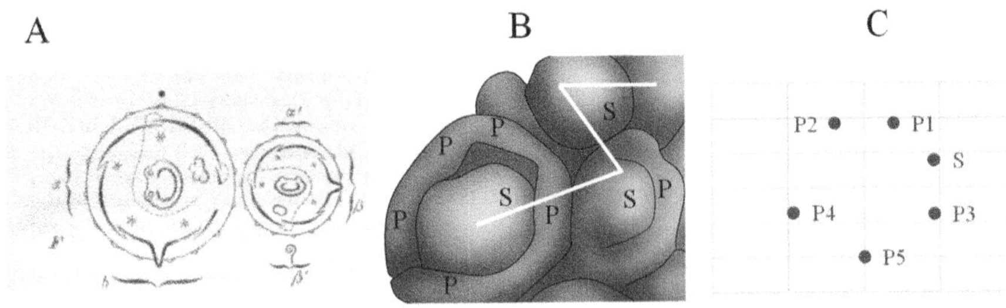

Figure 5.28. *(A) Inflorescence diagram of Centranthus ruber (Eichler 1875). The hook on the bottom represents the continuation of the cincinnus. (B) Flower meristems redrawn from Roels and Smets (1996). The development of the zigzagging scorpioid cyme is shown in white. (C) Model of the arrangement of the petal primordia. P = petal; S = single stamen*

Organ	Angle (°)	Distance from the center
Adaxial source	90	2.6
Bract	270	1
β	0	2
α	180	2

Table 5.15. *Modeling parameters for Centranthus ruber*

For modeling, petal initiation was considered to be directly dependent on the adaxial source of inhibition, the bract and the two bracteoles (before the initiation of the lateral flower) as per Eichler's diagram. The modeling places the single stamen alternating between two petals, which generates the asymmetry of the flower (Figure 5.28(C)).

5.9.5. *Dipsacoideae*

The flowers of Dipsacoideae are characterized by the presence of an epicalyx (Figure 5.29), which is considered to be a "key innovation" that spurred the adaptive radiation of Dipsacoideae in the Mediterranean Basin. The morphology of the epicalyx and the calyx has undergone several modifications within Dipsacoideae in response to different seed dispersal syndromes (Naghiloo and Claßen-Bockhoff 2017).

Most species in Scabioseae have an elaborate epicalyx with "wing-like" appendages and tend to have modified sepals (bristles). The combination of these characters facilitates wind dispersal. However, in most members of the Dipknautid clade (*Knautia*), the epicalyx is simple, and the calyx is reduced to a rim indicating that these structures are not directly involved in dispersal. Instead, the majority of species in this clade rely on a combination of rigid, often acuminate receptacular bracts at the base of the capitulum, and elastic stems that allow the projection of seeds over short distances (Naghiloo and Claßen-Bockhoff 2017).

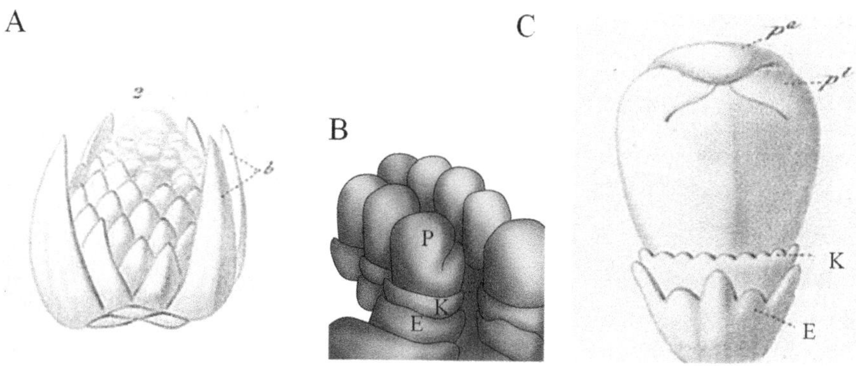

Figure 5.29. *The capitulum of Dipsacus laciniatus (Payer 1857). (B) Floral bud of D. sylvaticus (redrawn from an SEM from Roels and Smets 1996). (C) Flower development of D. laciniatus (Payer 1857). E = epicalyx; K = sepal; P = petal*

Knautia

The inflorescence of *Knautia arvensis* (Dipknautid clade) is a capitulum on which only the first-initiated, basal flowers are subtended by bracts, which arise successively in a **spiral** acropetal direction (Figure 5.30(A)). Unlike in *Scabiosa*, the upper flowers are not subtended by bracts (Figures 5.30(B) and 5.31(B)). During floral development, two median epicalyx primordia are initiated first, followed by the two transversely oriented ones. At approximately the same time, a ring-like meristem that represents the calyx is also visible slightly higher on the flower meristem, while a ring-like zone can be observed on top of the floral apex (Figures 5.30 (C) and (D)). Four broad petal primordia become visible on the ring-like zone. Finally, the calyx meristem differentiates into a number of protuberances that become bristles in later stages (Figure 5.30(D)). Then the outer sides of the epicalyx, calyx and corolla become hairy (Roels and Smets 1996).

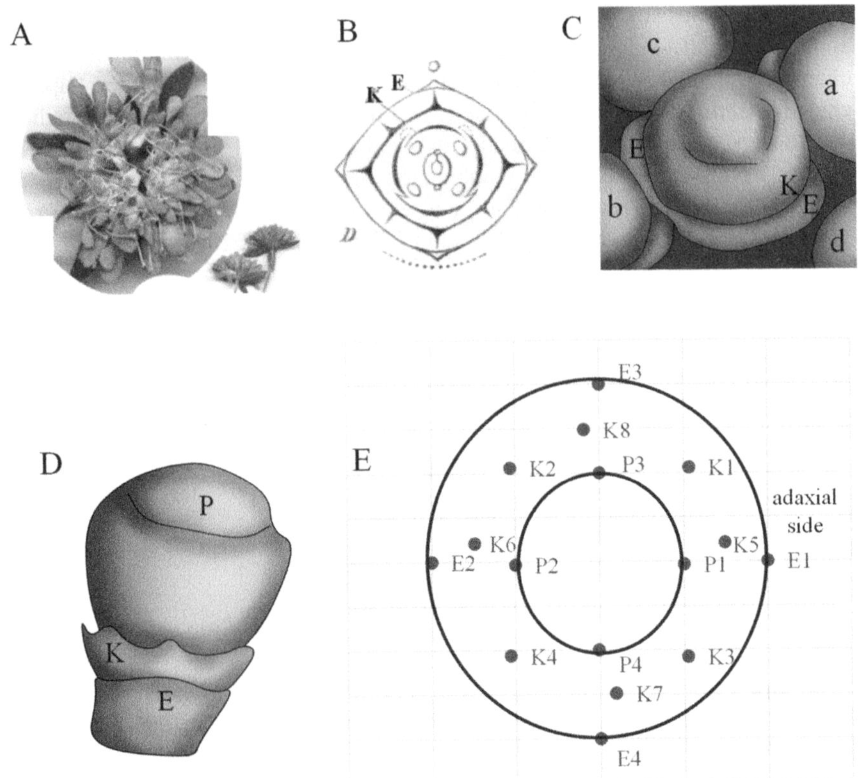

Figure 5.30. *(A) Knautia arvensis from Flore Virtuelle d'Orsay. (B) Floral diagram of K. arvensis (Eichler 1875). (C and D) Flower development (redrawn from Roels and Smets (1996)). (E) Model of the arrangement of the epicalyx and the perianth organs. E = epicalyx; K = sepal; P = petal*

The tetramerous epicalyx cycle is shaped by contact pressure from four neighboring FMs. The SEM of Roels and Smets (1996) shows that sepals K5 to K8 emerge from a ring-like zone after petal initiation, which allows the petals to alternate with sepals K1 to K4, then for sepals K5 to K8 to alternate with the first four sepals, in accordance with Eichler's diagram (Table 5.16 and Figure 5.30(B)).

Organ	Angle (°)	Distance from the center
a	45	3
b	225	3
c	135	3
d	315	3
E1-E4		2
K1–K4		1.5
P1–P4		1
K5–K8		1.5

Table 5.16. *Modeling parameters for Knautia arvensis. Note that a, b, c and d are the neighboring floral meristems (Figure 5.30(C))*

5.9.6. *Scabioseae*

Scabiosa

The merosity of the corolla tends to vary within Dipsacoideae. While tetramerous corollas are often found in Dipknautids, most Scabioseae are characterized by pentamerous corollas (reviewed by Naghiloo and Claßen-Bockhoff 2017).

In *Scabiosa ochroleuca*, the first sepal primordium appears on the adaxial side concomitantly with the formation of the epicalyx lobes, followed by the lateral and ventral sepals. The first petal primordium appears as a broad meristem on the adaxial side, which later splits into two petal primordia resulting in a pentamerous corolla (P1 and P2 as shown in Figure 5.31(D)). Bristle-like sepals enlarge quickly and cover the petals during development (Naghiloo and Claßen-Bockhoff 2017).

For modeling, pentamery is obtained by breaking the symmetry of the four parts of the epicalyx, one becoming less inhibitory than the other three (Table 5.17).

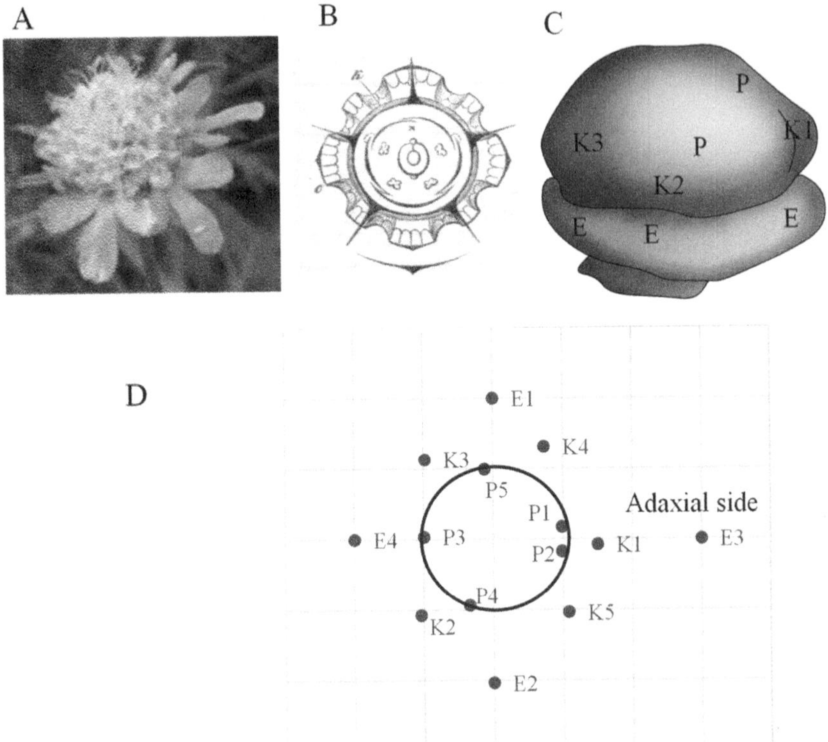

Figure 5.31. *(A) Scabiosa canescens from Flore Virtuelle d'Orsay. (B) Floral diagram of Scabiosa atropurpurea (Eichler 1875). (C) Flower development of Scabiosa ochroleuca (redrawn from Naghiloo and Claßen-Bockhoff (2017)). (D) Model using the data from Naghiloo and Claßen-Bockhoff (2017) (the adaxial side is on the right). E = epicalyx; K = sepal; P = petal*

Organ	Angle (°)	Distance from the center
E1	90	2
E2	270	2
E3	0	3
E4	180	2
K1–K5		1.5
P1–P5		1

Table 5.17. *Modeling parameters for Scabiosa*

6

Symmetry in Fabales

6.1. Introduction

The main subfamilies in Fabaceae (also known as Leguminosae) are Caesalpiniodeae and Faboideae (reviewed in Zhao et al. (2021)). Unlike most dicots, where the sepal on the dorsoventral axis is adaxial (K2), most species in these subfamilies have an abaxial sepal (K1) (e.g. Figure 6.3: *Cercis*; Figure 6.6: *Senna*; Figure 6.8: *Vicia*). Corolla aestivation is ascending cochlear in Caesalpiniodeae (Figure 6.6) and descending cochlear in Faboideae (Figure 6.8), the flowers of the two subfamilies being most often zygomorphic. Mimosoids are an exception among Caesalpiniodeae because they have actinomorphic flowers (Figure 6.7).

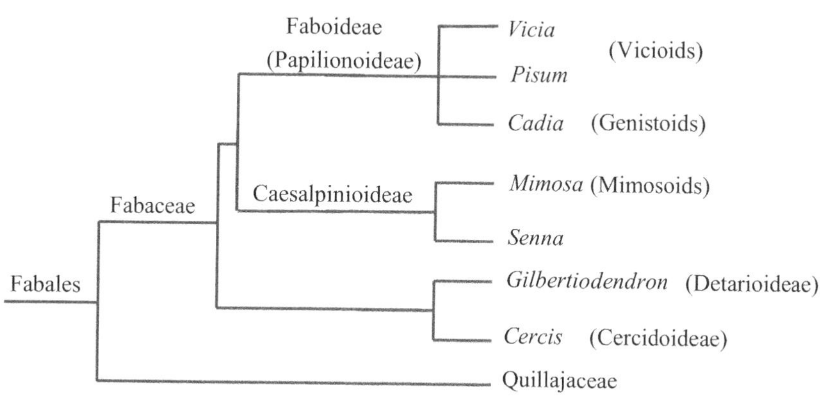

Figure 6.1. *Phylogenetic tree of the Fabales species described in this chapter (simplified from Zhao et al. (2021))*

Early diverging Fabales (Quillajaceae, Surianaceae and Polygalaceae) have dichasial cymes, actinomorphic flowers with one adaxial sepal (K2) and one abaxial petal, contorted petal aestivation, free petals that are similar in shape and free stamens. Their calices are initiated in a spiral sequence (Figure 6.2(A)). In Quillajaceae and Surianaceae, petals are initiated in a spiral pattern. In Caesalpinioideae, spiral, simultaneous, bidirectional and unidirectional patterns of petal initiation are observed. Unlike most other Fabales, Quillajaceae and Surianaceae have predominantly actinomorphic flowers (Bello et al. 2007).

An imbricate calyx is established in both Caesalpinoids and Mimosoids: sepals may overlap in the intermediate stages of development after initiation in a simultaneous, spiral sequence, or bidirectionally. This process is usually followed by an equalization process (section 1.3.2), where the sepals acquire a uniform size that allows the establishment of other types of calyx aestivation (e.g. valvate aestivation) at the end of floral development. An imbricate calyx at anthesis can be considered to be the result of the absence of equalization in the intermediate stages of flower development (De Barros et al. 2017).

Floral zygomorphy, considered to be a key feature correlated with higher net diversification rates in various lineages (Sargent 2004; Busch and Zachgo 2009; O'Meara et al. 2016; reviewed in Vamosi et al. (2018)), is of particular interest in Caesalpinioideae, where phylogenies show multiple switches from actinomorphy (polysymmetry) to zygomorphy (monosymmetry), and vice versa (Reyes et al. 2016). Floral zygomorphy is expressed in different ways in this group: through the loss or abortion of sepals, petals or stamens during development, or by the differentiation of organs within the same cycle (**heterochrony**: a process that can result in organ differentiation, the main one being differential development). In several members of Detarioideae, flowers are radially symmetrical at initiation and remain radially symmetrical at anthesis (e.g. Figure 6.4). In other Detarioideae, like *Gilbertiodendron*, flowers are radially symmetrical at initiation and become zygomorphic because of the differential development of organs within one or several cycles (Figure 6.5) (Bruneau et al. 2014).

6.2. Quillajaceae

Quillaja

Inflorescences of *Quillaja saponaria* are organized in dichasial (thyrsoid) cymes with a terminal flower flanked by lateral buds subtended by their respective bracts (Figure 6.2(A)). Following the sequential initiation of the bracteoles around each floral bud, the calyx is initiated in a spiral sequence, starting on the ventral side (K1), with the lateral sepals arising last (K4, K5; Figure 6.2(A)). As sepals initiate in the terminal flower, the lateral buds emerge laterally (arrows in Figure 6.2(A)).

Corolla initiation takes place after sepal initiation and is relatively fast. At anthesis, sepals are valvate and connate (Figure 6.2(B): Eichler 1878). The perianth displays near-radial symmetry (Bello et al. 2007).

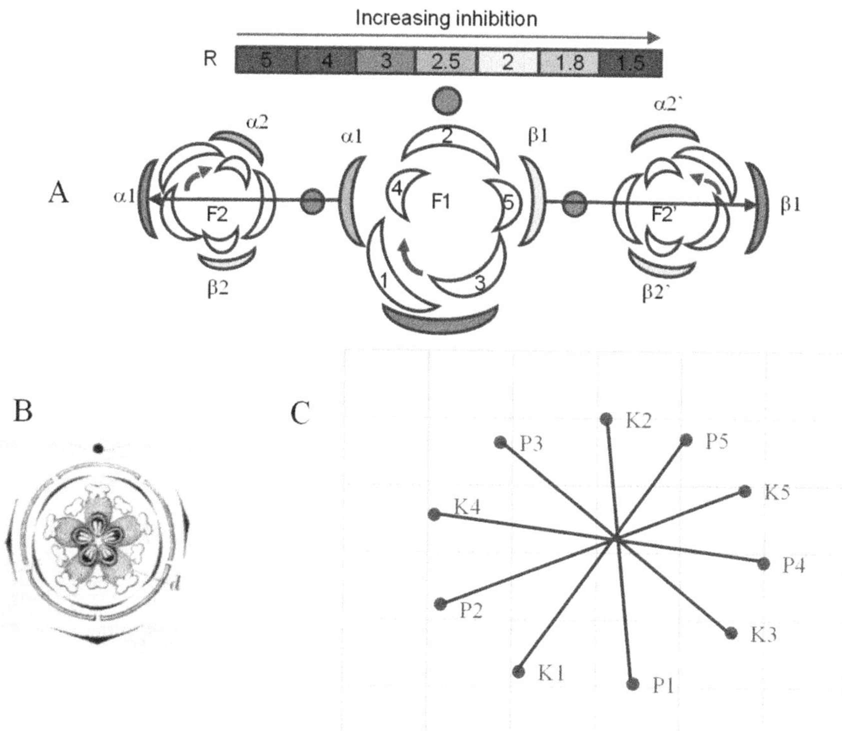

Figure 6.2. *(A) Diagram of a partial dichasial thyrsoid cyme in Quillaja saponaria (adapted from Bello et al. 2007). Blue circle = adaxial source for lateral flowers. (B) Floral diagram of Q. Brasiliensis at anthesis from Eichler (1878). (C) Model of the perianth organ initiation sequence in Q. saponaria. For a color version of this figure, see www.iste.co.uk/walch/phyllotaxis2.zip*

For lateral flowers (F2 and F2’: Figure 6.2(A)), the adaxial source of inhibition is the contact pressure exerted by the terminal flower (F1). Although the bracteoles are opposite (i.e. located on the transverse axis), their radial asymmetry allows the spiral initiation of the sepals (Table 6.1). The five axes of symmetry all intersect near the center of the flower, “the hub of the wheel” (Figure 6.2(C)).

Organ	Angle (°)	Distance from the center
Adaxial source	90	3
Bract	270	3
β	0	2
α	180	2.5

Table 6.1. *Modeling parameters for Quillaja saponaria*

6.3. Fabaceae

Fabaceae is the third-largest family of angiosperms with approximately 770 genera and 19,500 species. Some Fabaceae are of major economic importance for food and feed. The family is split into six subfamilies: Cercidoideae, Detarioideae, Duparquetioideae, Dialioideae, Caesalpinioideae (including the mimosoid clade) and Faboideae (LPWG 2013, 2017) (Figure 6.1).

6.3.1. *Cercidoideae*

Cercis

Cercis canadensis (Fabaceae) has spiral racemes, where acropetally initiated bracts subtend individual flowers. Paired bracteoles are initiated in succession from the floral apex (Figure 6.3(A)). Sepals are chronologically initiated in a spiral sequence, starting with the single ventral sepal (K1), the first dorsal sepal (K2), two lateral sepals (K3 and K4) and the second dorsal sepal (K5) (Figure 6.3(A)). Petals are initiated unidirectionally from the abaxial side. "Papilionate" features appear only after the flower bud is over half its final size: they include the development of a calyx tube and morphological differences between petals. Aestivation is ascending cochlear (with the wings [P3 and P4] overlapping the vexillum [P5]), rather than descending cochlear (with the vexillum overlapping the wings) as in Faboideae (Figure 6.3(B) and 6.3(C); Tucker 2002a).

For modeling, the two bracteoles were considered to be angularly and radially symmetrical: the angle between the bracteoles (circa 137°; see Eichler diagram: Figure 6.3(C)) is congruent with their spiral initiation. Given these parameters, the initiation of the calyx is spiral. Bilateral symmetry of the first four petals is forced; otherwise, petal P2 would be in a lateral (instead of a ventral) position and petal initiation would therefore not be unidirectional. Zygomorphy in *C. canadensis* is constitutional.

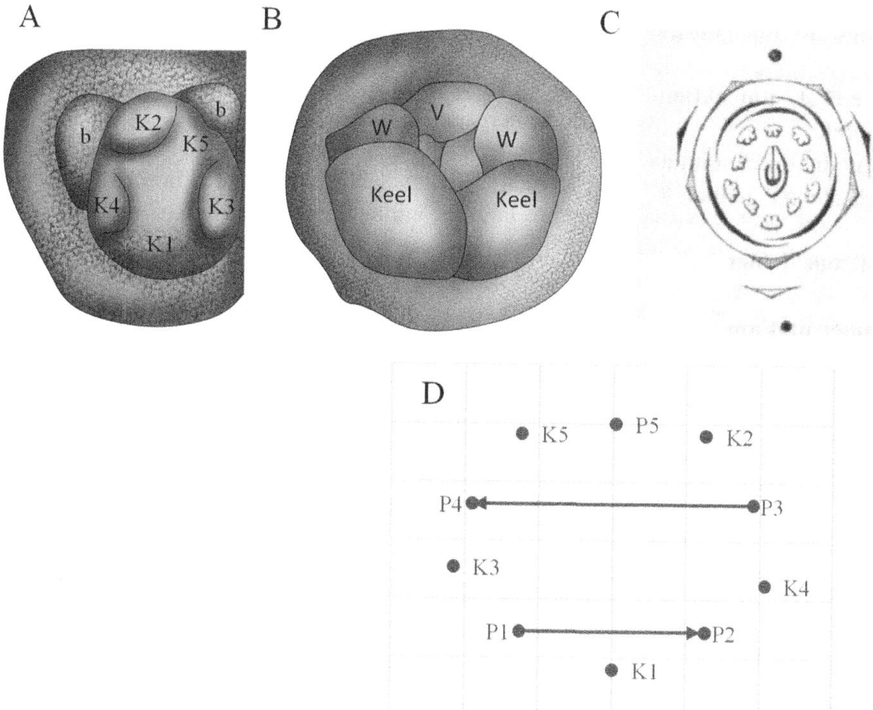

Figure 6.3. *(A) Calyx initiation in Cercis canadensis. The five sepals are initiated in a spiral sequence. (B) Ascending cochleate petal aestivation in C. canadensis (from Tucker 2003). (C) Floral diagram of Cercis siliquastrum (Judas tree) from Eichler (1878). (D) Model of the perianth organ initiation sequence in C. canadensis. b = bracteole; V = vexillum; W = wing. For a color version of this figure, see www.iste.co.uk/walch/phyllotaxis2.zip*

Organ	Angle (°)	Distance from the center
Adaxial source	90	2
Bract	270	4
a	20	2
b	160	2

Table 6.2. *Modeling parameters for Cercis canadensis*

6.3.2. *Detarioideae*

Isoberlinia

Bracts on the inflorescence meristem of *Isoberlinia angolensis* (Detarioideae) are initiated in a spiral sequence. An oval floral apex (wider along the transverse axis) is formed in the axil of each bract. Two bracteoles are initiated successively by the floral apex. The bracteoles are large at initiation, and rapidly encircle 90% of the circumference of the floral meristem, so that the floral apex is omega-shaped, tapered adaxially and truncated abaxially. The five sepals have a spiral initiation sequence, with the first sepal (K1) initiated abaxially. Petals are initiated near-simultaneously. Flowers are actinomorphic (Tucker 2002b).

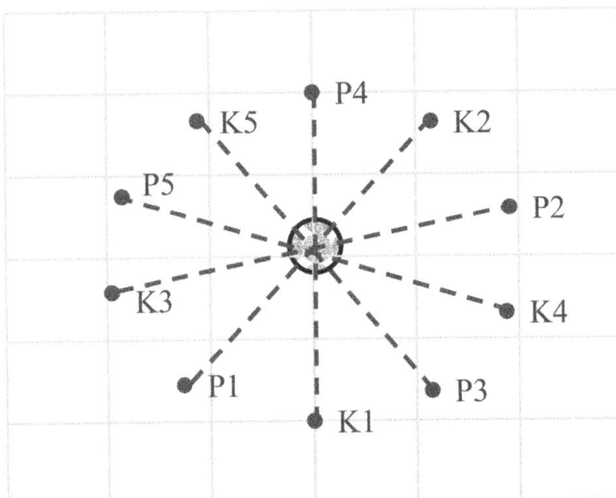

Figure 6.4. *Model of the actinomorphic flower of I. angolensis with five axes of symmetry intersecting near the center of the flower*

Organ	Angle (°)	Distance from the center
Adaxial source	90	2
B	270	4
a	20	2
b	160	2

Table 6.3. *Modeling parameters for Isoberlinia angolensis*

For modeling, the two bracteoles were considered to be angularly and radially symmetrical: the angle between the bracteoles (circa 137°) is congruent with their spiral initiation. With these parameters, sepal initiation is spiral. The bilateral symmetry of the petals is not forced, the radial symmetry of the perianth being the consequence of the symmetry of the bracteoles ($R_a = R_b$) and the pentamery being the consequence of the spiral position of these bracteoles (the divergence angle is close to the golden angle = 137°) (Figure 6.4).

Gilbertiodendron

Gilbertiodendron (Detarioideae) is zygomorphic at anthesis. Bract primordia are initiated on the apical meristem of the racemose inflorescence in a spiral sequence. An oval floral apex (wider along the dorsoventral axis) is initiated in the axil of each bract. Paired bracteoles are initiated from the floral apex in rapid succession and grow rapidly to surround the floral apex, before and during floral organogenesis. After bracteole initiation, the floral apex is narrow and tapered adaxially and abaxially. Sepal initiation is sequential along a "2/5" quincuncial spiral. The first sepal primordium (K1) is initiated on the abaxial side. The base of the sepal primordia becomes marginally confluent to form a shallowly lobed calyx tube. Petal initiation is unidirectional, with the first four petals appearing simultaneously and the adaxial petal lagging. The five petal primordia remain small and uniform in size through midstage development. At anthesis, the vexillary (adaxial) petal is much larger than the others; it is fan-like, with two lateral lobes and clawed. The four other petals are small and linear-lanceolate (Tucker 2002b) (Figure 6.5).

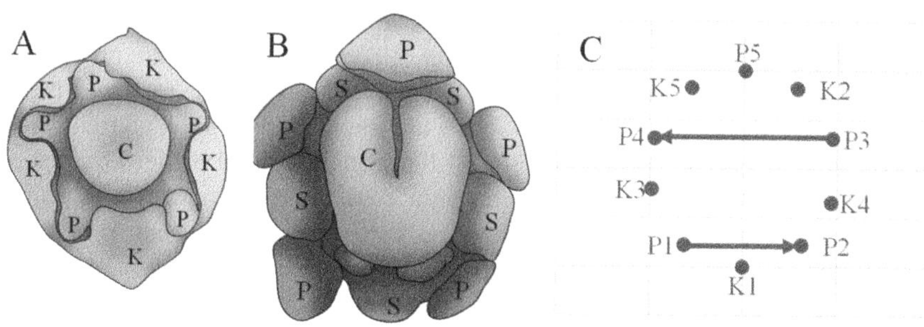

Figure 6.5. *Floral bud development in Gilbertiodendron klainei: (A) Sepal, petal and carpel primordia have been initiated. (B) Floral bud with all petals expanding and an open carpel cleft. Three of the antesepalous stamen primordia (the three ventral stamens) exceed the other two in size (Tucker 2002b). (C) Model of the perianth organ initiation sequence in G. klainei. K = sepal; P =petal; S = stamen; C = carpel. For a color version of this figure, see www.iste.co.uk/walch/phyllotaxis2.zip*

Organ	Angle (°)	Distance from the center
Adaxial source	90	2
B	270	4
a	20	2
b	160	2

Table 6.4. *Modeling parameters for Gilbertiodendron klainei*

For modeling, the two bracteoles were considered to be angularly and radially symmetrical (Table 6.4: 90±70°): the angle between the bracteoles (circa 137°) is congruent with their spiral initiation. However, with these model parameters, the initiation of the calyx is spiral. The bilateral symmetry (zygomorphy) of the first four petals is forced (Figure 6.5(C)); otherwise, petal P2 would be in a dorsal (in place of P3) instead of a ventral position and initiation would therefore not be unidirectional (ascending cochlear aestivation). The difference with *Isoberlinia* is the unidirectional initiation of petals, which is the mark of zygomorphy.

6.3.3. *Caesalpinioideae*

6.3.3.1. *Cassioideae*

Cassia javanica

The terminal inflorescences of *Cassia javanica* are racemes with spirally arranged bracts, each subtending a flower. Each floral apex first produces two opposite bracteoles in succession, which rapidly become equal in size. The floral apex then becomes circular, and the sepals are initiated in a spiral, starting with the abaxial sepal (K1), positioned slightly off the median sagittal plane. The first sepal enlarges, becoming essentially median as a result. The second sepal is initiated dorsally and the third laterally. The five petals alternate the sepals, arise over a brief period of time and their initiation sequence appears unidirectional. The calyx and corolla become imbricate. The abaxial sepal remains the largest and encloses the others. The petals are alike in size and shape before aestivation (Tucker 1996).

For modeling, two bracteoles were placed symmetrically (angularly and radially; Table 6.5: 90±70°): the calyx is thus quincuncial. The bilateral symmetry of petals P1 and P2, and P3 and P4, respectively, was forced. Their arrangement at initiation is thus ascending cochlear (Figure 6.6).

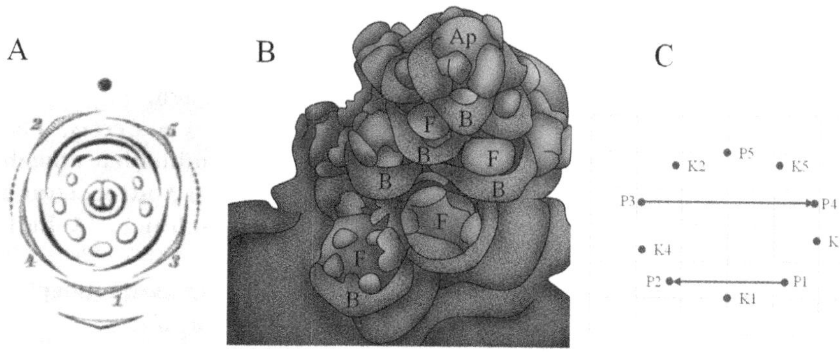

Figure 6.6. *(A) Floral diagram of Senna occidentalis (syn. Cassia caroliniana) with cryptic bracteoles (Eichler 1878). (B) Developing racemose inflorescence of S. didymobotrya. Flowers are initiated sequentially and acropetally (with the oldest, first formed flowers at base, and successively younger ones above) in the axil of the bracts (Tucker 1996). (C) Model of the perianth organ initiation sequence in Cassia with bracteole. Ap = inflorescence apex; B = bract; F = flower. For a color version of this figure, see www.iste.co.uk/walch/phyllotaxis2.zip*

Organ	Angle (°)	Distance from the center
Adaxial source	90	2
Bract	270	4
b	20	2
a	161	2

Table 6.5. *Modeling parameters for Cassia javanica*

Senna didymobotrya

Senna didymobotrya has spiral phyllotaxy, axillary inflorescences and zygomorphic yellow or orange flowers. The inflorescence structure is racemose, with one flower in the axil of each bract. Bracteoles are absent. Sepals are initiated in a spiral, starting in the abaxial position (K1); the second sepal (K2) is dorsal and off-median, and the third and fourth sepals are lateral. The fifth sepal is adaxial and next to K2; the timing of the initiation of K3–K5 overlaps with that of the petals. Petals are initiated near-simultaneously. The first petals are the two abaxial petals and one lateral petal. The other lateral petal and the adaxial petal appear later. Once all five petals have been initiated, they rapidly become equal in size. Buds are circular in cross-section; zygomorphy becomes accentuated by allometric growth, differential organ expansion and organ inflexion as the flower opens (Tucker 1996).

The initiation of the calyx and corolla of *Senna didymobotrya* is identical to that of *Cassia javanica*, despite the absence of bracteoles: the forces shaping the perianth are not due to biochemical activity but to contact pressure from the bracts subtending the flowers above each developing flower in the inflorescence (Figure 6.6(B)).

6.3.3.2. *Mimosoids*

The Mimosoid clade differs from other Fabaceae by a number of structural characteristics: the absence of bracteoles, radial floral symmetry, which is usually associated with globular inflorescences, and the formation of sepal and petal tubes. In Mimosoids, there is much variation in the initiation sequence of the sepals (spiral, simultaneous), calyx development being generally followed by an equalization process that leads to valvate aestivation (De Barros et al. 2017). Petals are generally very small, and this is correlated with the presence of long, showy filaments (polyandrous flowers) (Ronse de Craene 2022).

Mimosa

Mimosoid (Fabaceae) inflorescences are usually racemous (Figure 6.7(A), *Mimosa pudica*), bearing 90–100 spirally arranged flowers. Some of these inflorescences share an unusual developmental feature: the synchronous development of all flowers in an inflorescence. As in the racemes of the other subfamilies of Fabaceae, flowers are initiated in an acropetal successive order; however, floral bud development is paused until all flowers have been initiated in that inflorescence. All flowers then undergo a synchronized initiation of sepals, petals, etc. As a result, all flowers in an inflorescence are at the same developmental stage. Synchronous development in mimosoid inflorescences contrasts with the successive development found in inflorescences of the other Fabaceae subfamilies (Tucker 2003).

Flowers are radially symmetrical (actinomorphic) in nearly all taxa in the mimosoid subfamily. Radially symmetrical mimosoid flowers usually have four or five organs in each cycle, and all members of a cycle are alike.

A spiral raceme generates an actinomorphic flower (Figure 6.7 and Table 6.6). The modeling takes into account the six neighboring primordia of each flower on the inflorescence meristem.

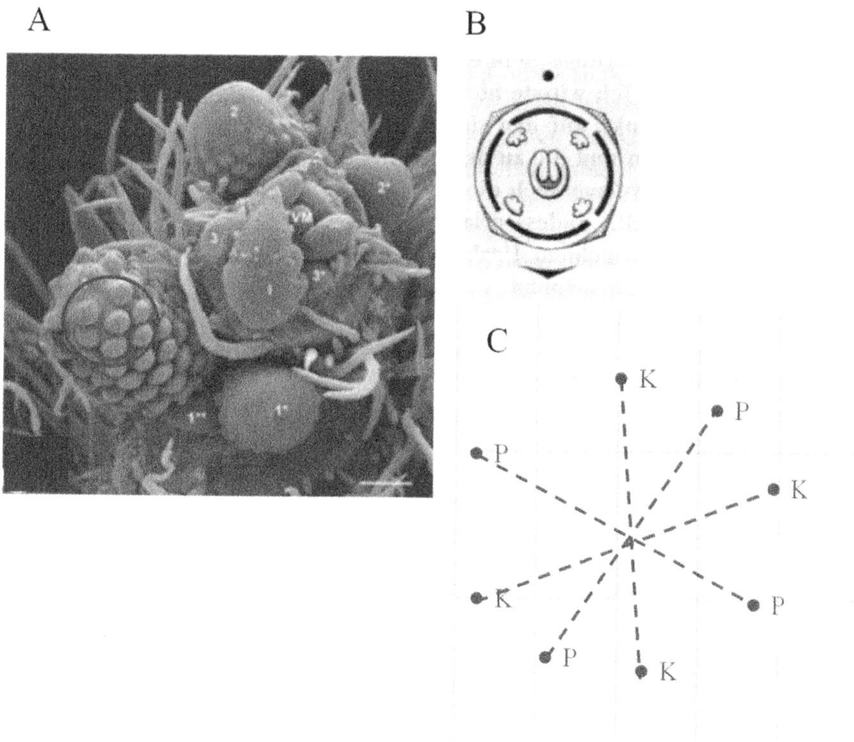

Figure 6.7. *(A) Mimosa pudica inflorescence showing neighboring primordia: open flowering shoot system (vegetative meristem) with heads at different developmental stages (1–3; same number and asterisks indicate accessory floral units in same leaf axil). This is reproduced with permission from Claßen-Bockhoff and Bull-Hereñu (2013). (B) Floral diagram of tetramerous M. pudica (Eichler 1878). (C) Model of the perianth organ initiation sequence in tetramerous Mimosa pudica. For a color version of this figure, see www.iste.co.uk/walch/phyllotaxis2.zip*

Organ	Angle (°)	Distance from the center
F1	85	2.5
F2	115	2
F3	145	2.5
F4	65	2.5
F5	295	2
F6	325	2.5

Table 6.6. *Mimosa modeling parameters (neighboring primordia from Figure 6.7(A))*

6.3.4. *Faboideae*

Faboideae (Papilionoideae) have zygomorphic flowers: the adaxial petal is the flag/banner, the two lateral petals are the wings and the two abaxial petals form a keel protecting the fertile organs. At anthesis, corolla aestivation is descending cochlear, unlike in Caesalpinioideae, where it is ascending cochlear (Figure 6.8; Tucker 2003).

Most flowers in Faboideae have a single pentamerous cycle of sepals, a single pentamerous cycle of petals, two cycles of stamens and a single carpel (Tucker 2003). Furthermore, floral organs are normally initiated in the following order: sepals, carpel, petals and stamens. Within each cycle, the organs are initiated unidirectionally from the abaxial to adaxial side (Dong et al. 2005); however, at anthesis, the corolla is descending cochlear because of differential organ development (Figure 6.8).

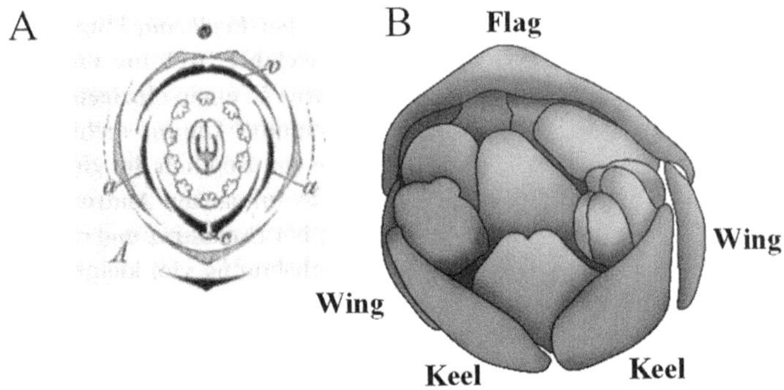

Figure 6.8. *(A) Floral diagrams of Vicia faba (Faboideae) with cryptic bracteoles. v = "vexillum" (bilobed flag); a = "alea" (wing); c = "carina" (keel). (B) Zygomorphic flower: Genista tinctoria (Faboideae; redrawn from Tucker 2003). For a color version of this figure, see www.iste.co.uk/walch/phyllotaxis2.zip*

The floral structure is the same with or without bracteoles, these being cryptic in the latter (Unterdrückt, according to Eichler).

Pisum

In *Pisum sativum* (Faboideae), at the start of the reproductive phase, each shoot becomes a primary inflorescence (I1; not shown) and produces secondary inflorescence meristems (I2: Figures 6.9 (A) and (B)). No indication of a terminal flower on the second-order axis is found. The second-order inflorescence apex produces several successive leaves, each subtending a single-flowered branch (third-order axis: I3). Four to eight third-order axillary buds are produced by the second-order inflorescence meristem before it ceases activity. Each third-order axillary inflorescence meristem produces a floral apex (Fn; Tucker 1989).

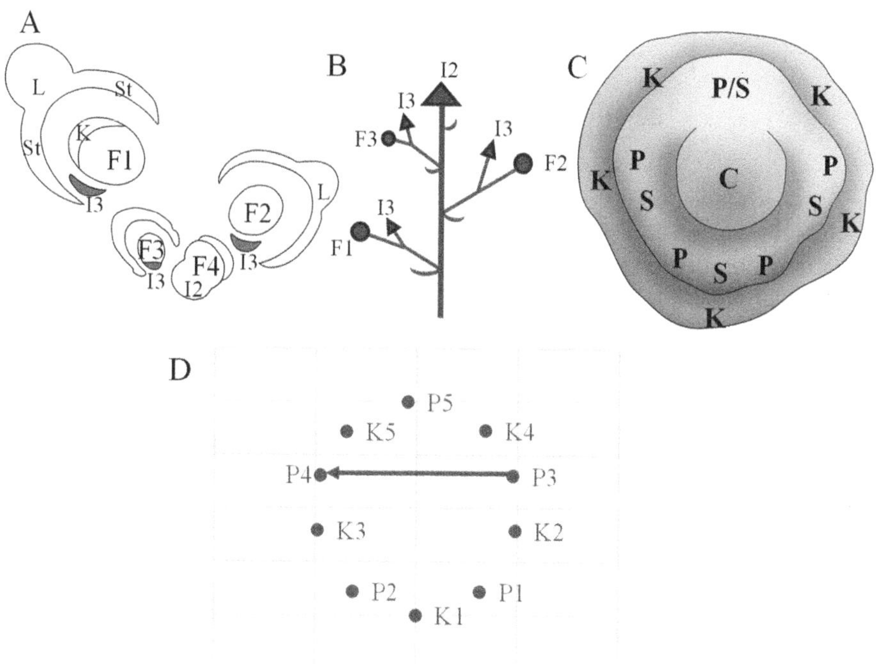

Figure 6.9. *Pisum sativum. (A) Transverse section of an axillary inflorescence. F1–F4: flower meristems; I2: second-order inflorescence apex; I3: third-order inflorescence apex; L = leaf; K = sepal; St = stipule (from Tucker 1989). (B) Inflorescence diagram of P. sativum (from Tucker 1989). The second-order inflorescence apex is represented by a large triangle. Each axillary shoot has one terminal flower (circle) and a third-order inflorescence apex (smaller triangle). (C) Floral meristem of P. sativum (from Tucker 2003): petals and stamens are initiated from common primordia and the timing of the initiation of different cycles overlaps. (D) Model of the perianth organ initiation*

Floral meristems are generated from I2. The floral ontogeny of *P. sativum* shows a succession of sepals, petals plus carpel, antesepalous stamens and antepetalous stamens. Within each cycle, organs arise unidirectionally, beginning, as in most Faboideae, on the abaxial side of the flower (K1). Unusual features include the four common primordia, which precede initiation of discrete petal and antesepalous stamen primordia, and the marked overlap of organ initiations between cycles, which are usually separately initiated (Tucker 1989) (Figure 6.9(C)).

Organ	Angle (°)	Distance from the center
Adaxial source	90	1.4
Leaf	270	4
Stipule	20	2
Stipule	160	2

Table 6.7. *Modeling parameters for Pisum sativum*

In our model, one stipule takes on the role of two bracteoles with regards to the inhibition potential. The floral meristems are circular, which sets them apart from the oval floral meristems of *Antirrhinum*, which are deformed by contact pressure within large racemes. The early initiation of the carpel does not modify the abaxial position of the first petals. The inhibition from the adaxial source remains predominant.

In *Pisum*, *CYC*-like genes (*CYC1*, *K* and *LST1*) originating from the duplication of a *TCP* gene during Faboideae diversification control the identity of the lateral and dorsal floral organs (Figure 1.11(A); Wang et al. 2008).

Cadia

Cadia purpurea (Genistoid clade) is a Faboideae with radial symmetry, a campanulate calyx (tube with five triangular lobes), unstable petal aestivation and free organs, all unusual features among Faboideae. Flowers are usually solitary or, rarely, in few-flowered racemes. The order of organ initiation is unidirectional in each cycle, and carpel initiation takes place directly after petal initiation. The sepal primordia enlarge uniformly, forming a five-lobed calyx tube that is relatively thick and confluent by the early petal stage. In bud, the sepal lobes remain valvate and grow as fast as the calyx tube, so that the free sepal lobes are approximately equal in size and shape at maturity. When the adjacent growing petal margins of *Cadia* meet, chance determines which petal becomes positioned inside the neighboring petal. Hence, the pattern of petal aestivation is random (Tucker 2002c).

In typical Faboideae, organogenesis is dorsoventrally asymmetric: after a phase of uniform organ growth, at an advanced stage of floral ontogeny, unequal development occurs along the dorsoventral axis (Tucker 2003). By contrast, in *Cadia purpurea*, floral organs continue to develop equally until maturity (Tucker 2002c).

Cadia purpurea has reverted to radially symmetrical flowers. Citerne et al. (2006) investigated the genetic basis of this morphological reversal. Orthologues of the floral symmetry gene (*CYC*) demarcate the adaxial (dorsal) region of the flower in typical Faboideae (e.g. *Pisum*; Figure 1.11). In the corolla of *C. purpurea*, *LegCYC1A* is expressed only in the dorsal petal, and its level of expression appears very weak. *LegCYC1B*, however, is expressed in all petals, showing an expansion of the expression domain of this gene that is consistent with the radial phenotype of the corolla (Figure 6.10).

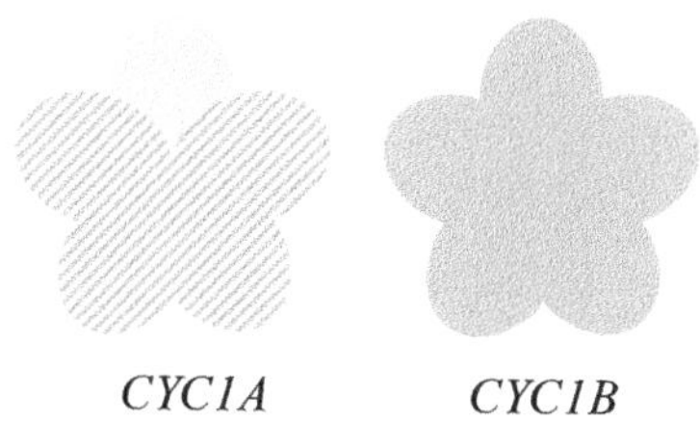

Figure 6.10. *Cadia purpurea: expression of LegCYC1A and LegCYC1B in the corolla of C. Purpurea (from Citerne et al. 2006). Blue = expression of CYC-like genes in the floral meristem; striped = weak expression. For a color version of this figure, see www.iste.co.uk/walch/phyllotaxis2.zip*

7

Symmetry in Monocots

7.1. Introduction

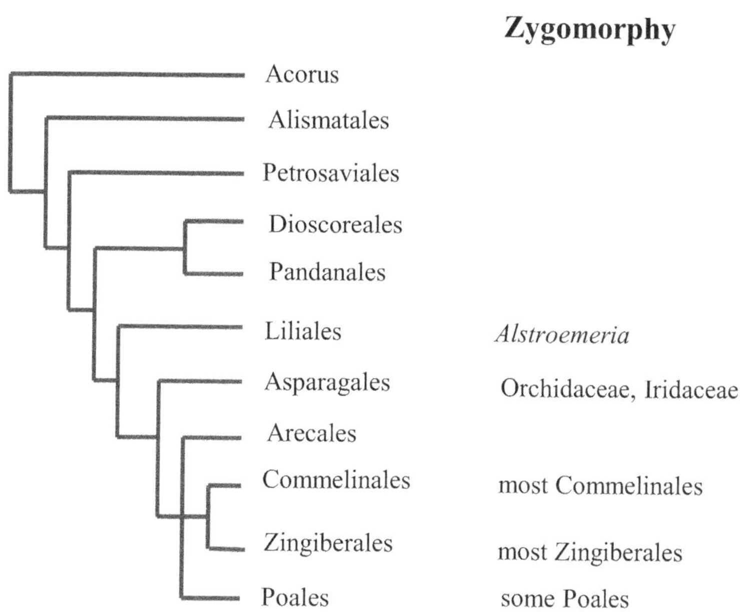

Figure 7.1. *Phylogenetic tree of monocots according to APG IV (2016)*

For a color version of all the figures in this chapter, see www.iste.co.uk/walch/phyllotaxis2.zip.

Unlike eudicots, flowers of monocots generally do not have a bracteole (Figure 7.2(A)) or have only one (Figures 7.2(B) and (C)); they rarely have two. They frequently show little morphological distinction between sepals and petals; hence, in monocots, these structures are often termed outer tepal (OT) and inner tepal (IT), respectively, with each of these cycles counting three pieces.

When phyllotaxy on the primary inflorescence axis is distichous, the bracteole is usually in an adaxial position (Figure 7.2(B)). If phyllotaxy is spiral on the primary inflorescence axis, the position of the bracteole on the pedicel is more or less transverse (Figure 7.2(C); e.g. Petrosaviaceae, Liliaceae, Dioscoreaceae and Nartheciaceae) (Remizowa et al. 2013).

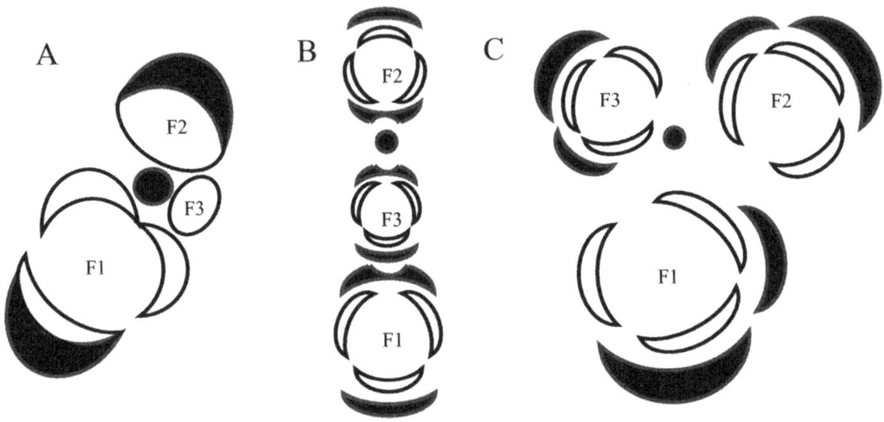

Figure 7.2. *(A) Partial inflorescence without a bracteole (illustrated by Telipogon falcatus, Orchidaceae, from Pabon-Mora and Gonzalez (2008)). (B) Partial distichous inflorescence with an adaxial bracteole. (C) Partial spiral inflorescence with a lateral bracteole (redrawn from Remizowa et al. (2013)). Black: the inflorescence axis, bracts and bracteoles; white: outer tepals; F: floral meristems numbered according to their chronological order of initiation*

7.2. Acorales

Acorus

Acorales consist of the single family Acoraceae and one genus *Acorus*. *Acorus* appears with strong support as sister to all other monocots (Figure 7.1) (Soltis et al. 2018).

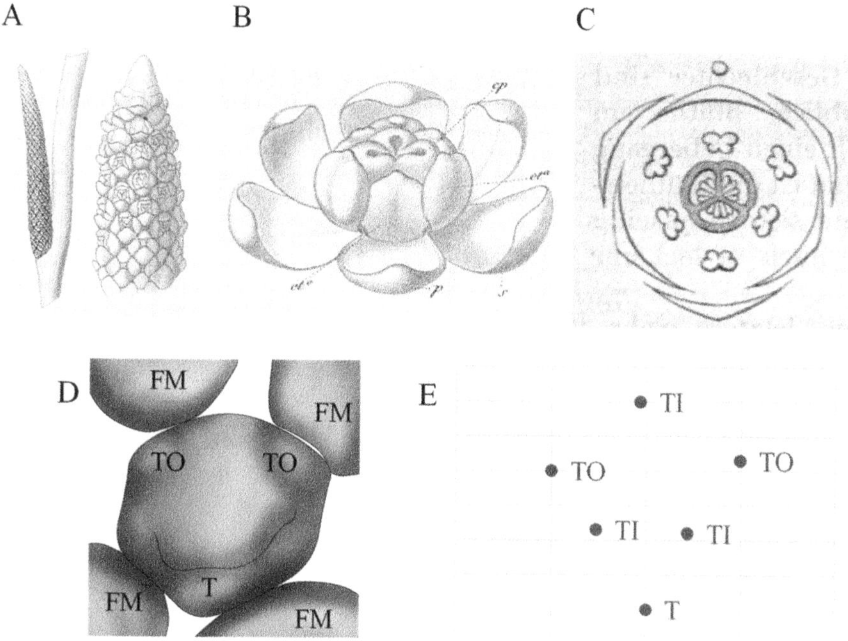

Figure 7.3. *Acorus calamus. (A) Inflorescence (spadix) and (B) flower from Payer (1857). (C) Floral diagram (Eichler 1875). (D) SEM of young flower showing peripheral floral meristems (FM). T = the bractlike tepal. TO = lateral mounds indicating adaxial tepals of outer perianth cycle (from Buzgo and Endress (2000)). (E) Model of perianth, according to Table 7.1*

The flowers are densely arranged in an elongated spadix without bracts (Figure 7.3(A)). Due to the close packing, neighboring flowers touch each other. However, the meristem is not flattened like that of *Antirrhinum*. For modeling, we use contact pressure from four peripheral floral meristems and no inhibition from the inflorescence axis (Figure 7.3(D)). The model fits the SEM and the floral diagram (Figure 7.3(E)).

Organ	Angle (°)	Distance to the center
FM	60	2
FM	120	2
FM	225	3
FM	315	3

Table 7.1. *Acorus calamus modeling parameters. FM: the four peripheral floral meristems, as shown in Figure 7.3(D)*

7.3. Alismatales

Following *Acorus*, Alismatales are sister to the remaining monocots, and the remaining monocots form a well-supported clade. For a small clade, floral diversity is exceptionally high in Alismatales (Soltis et al. 2018).

7.3.1. *Melanthiaceae*

Tofieldia

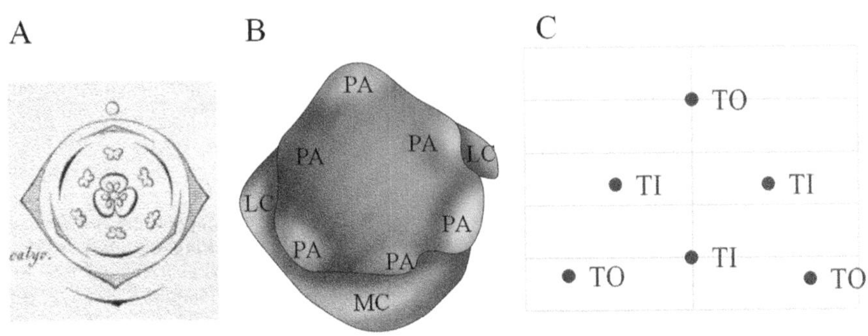

Figure 7.4. *(A) Floral diagram of Tofieldia calyculata (Eichler 1875). (B) SEM of T. pusilla: initiation of primordia common to tepals and stamens. PA = perianth plus androecium (Remizowa et al. 2006). MC = median calyculus scale; LC = lateral calyculus scale primordium. (C) Model. TO = outer tepal; TI = inner tepal*

Organ	Angle (°)	Distance to the center
LC_1	0	2
LC_2	180	2
MC	270	2

Table 7.2. *Tofieldia pusilla modeling parameters. LC1, LC2 and MC are the three peripheral calyculi, two of which are transverse*

All *Tofieldia* (Melanthiaceae) possess an unusual structure and are termed as calyculus (Figures 7.4(A) and (B)). This structure consists of three free or connate phyllomes alternating with the outer tepals.

As a result of the presence of a calyculus, *Tofieldia* differ strongly in floral orientation from most other monocots that possess bracteate racemose inflorescences, because the median outer tepal is adaxial (Remizowa et al. 2012). This is because the inhibition by the abaxial calyculus is stronger than that of a "simple" bract.

The flower is actinomorphic, and the aestivation is valvate. The model fits the SEM of *T. calyculata* (Figure 7.4).

7.3.2. *Alismataceae*

Alisma

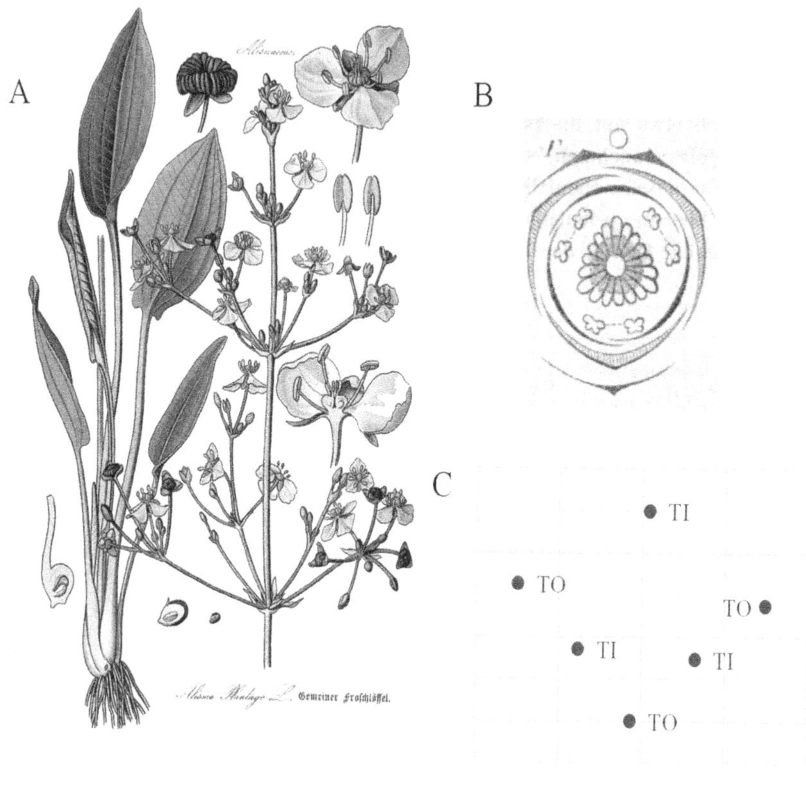

Figure 7.5. *Alisma plantago. (A) Mature plant (Wikipedia Commons: in the public domain). (B) Floral diagram: flower with a bract and an adaxial bikeeled bracteole (Eichler 1875). (C) Model of A. Plantago*

Organ	Angle (°)	Distance to the center
b1	80	3
b2	100	3
B	270	3

Table 7.3. *Alisma plantago modeling parameters*

For modeling, the bikeeled bracteole is split into two symmetrical dorsal sources of inhibition (b1 and b2). The initiation of the tepals is then spiral, and the flower becomes actinomorphic at anthesis, with six stamens. The model fits the floral diagram from Eichler (Figures 7.5(B) and (C)).

7.3.3. *Hydrocharitaceae*

Elodea

Hydrocharitaceae are submersed aquatics, some of which are marine (e.g. *Halophila*, *Enhalus* and *Thalassia*). The two bracteoles of *Elodea canadensis* form a spathe (Soltis et al. 2018) (Figure 7.6(A)).

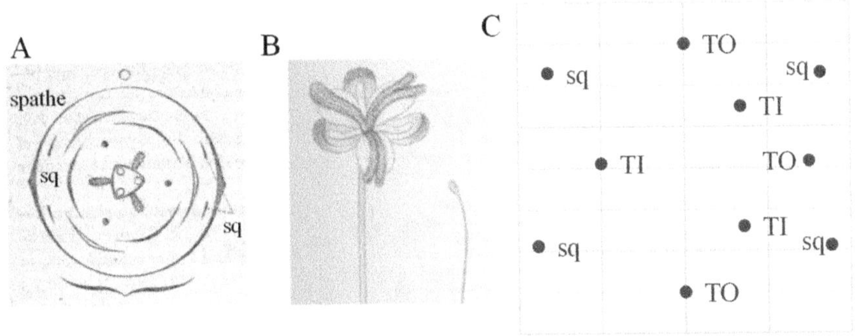

Figure 7.6. *Elodea canadensis. (A) Floral diagram (Eichler 1875). (B) Mature flower from Carl Axel Magnus Lindman. (C) Model. sq = squamulae intravaginales or stipule; TO = outer tepal; TI = inner tepal*

The modeling requires taking into account the intravaginal scale-like squamules (sq) located between the spathe (b1 and b2). Bilateral symmetry is broken from the first tepal. The model fits the floral diagram from Eichler (Figures 7.6(B) and (C)).

Organ	Angle (°)	Distance to the center
Adaxial source	90	1.2
B	270	1.2
b1	0	1.9
b2	180	1.9
		Plastochron ratio
Sq		1
Sq		1
Sq		1
Sq		2
TO		1
TO		1
TO		1.5
TI		1
TI		1
TI		1

Table 7.4. *Elodea canadensis modeling parameters*

7.4. Dioscoreales

7.4.1. *Nartheciaceae*

Metanarthecium

The flowers of *Metanarthecium luteo-viride* (Nartheciaceae) are arranged in long racemes, subtended by long, linear bracts and have a small transversal bracteole in the upper part of the pedicel. The first tepal is initiated strictly opposite the bracteole and the outer tepals arise in a spiral sequence. The flowers are actinomorphic, trimerous, bisexual, with six basally connate tepals in two cycles, six stamens in two cycles and three (sometimes four) almost united carpels forming a cone-shaped gynoecium (Remizowa et al. 2008).

If the bracteole is in a transverse position, TO1 is transverse and opposite the bracteole (Figure 7.7). The model fits the SEM from Remizowa et al. (2008) (Figures 7.7(B) and (C)).

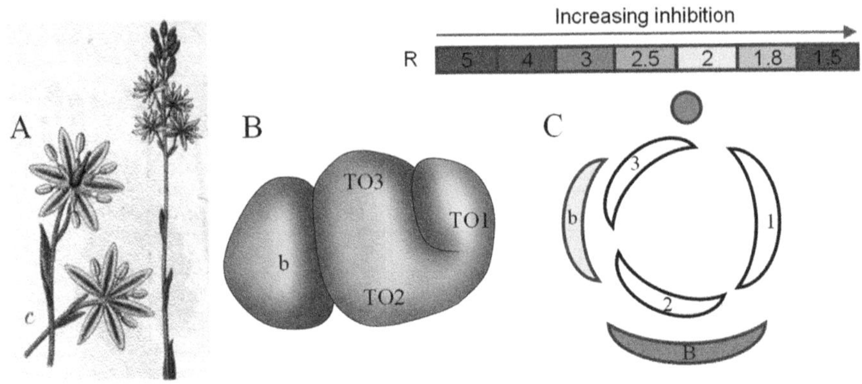

Figure 7.7. *(A) Narthecium ossifragum from Johann Georg Sturm (1796). (B) Initiation of outer tepals of Metanarthecium luteo-viride (Nartheciaceae) from Remizowa et al. (2008). (C) Model of meristem with transverse bracteole and one transverse sepal. The color represents the apparent distances from the sources of inhibition to the apex (R), the intensity of the inhibition being all the more important, as the distance from the organ to the apex is low (see Chapter 1). b = bracteole; TO = outer tepal*

7.4.2. *Dioscoreaceae*

Dioscorea

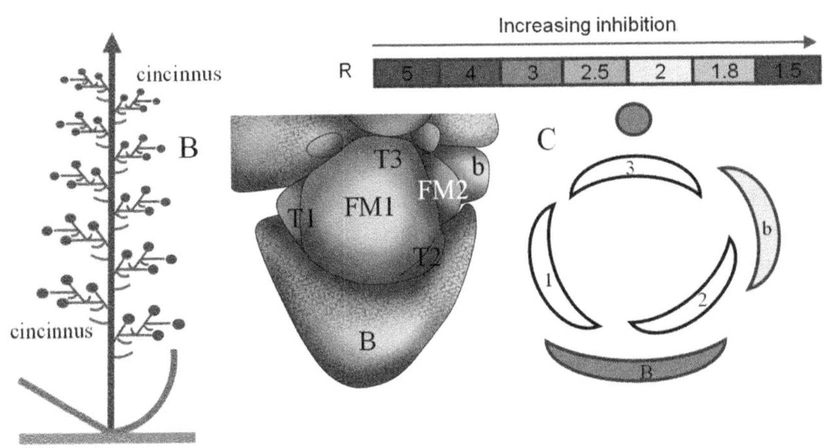

Figure 7.8. *(A) Inflorescence showing cincinni. (B) Dioscorea tokoro: the bracteole and all three outer tepals are initiated; young floral meristem (FM2) is visible in the axil of the bracteole (Remizowa et al. 2010). (C) Model. B = bract; b = bracteole; T = tepal. The color represents the apparent distances from the sources of inhibition to the apex (R), the intensity of the inhibition being all the more important as the distance from the organ to the apex is low (see Chapter 1)*

Inflorescence and early flower development in the East Asian *Dioscorea tokoro* were investigated by Remizowa et al. (2010) using scanning electron microscopy (SEM). The inflorescence is a thyrse (racemes + cymes; Walch and Blaise 2023) (Figure 7.8 (A)). Lateral units of thyrses are cincinni. The first tepal is inserted opposite the bracteole. The model fits the SEM from Remizowa et al. (2010) (Figures 7.8(B) and (C)).

The F2 flower primordium is initiated between F1 and its bracteole, even before the formation of the inner tepals of F1 (Figure 7.8(B)).

7.5. Liliales

7.5.1. *Liliaceae*

Lilium

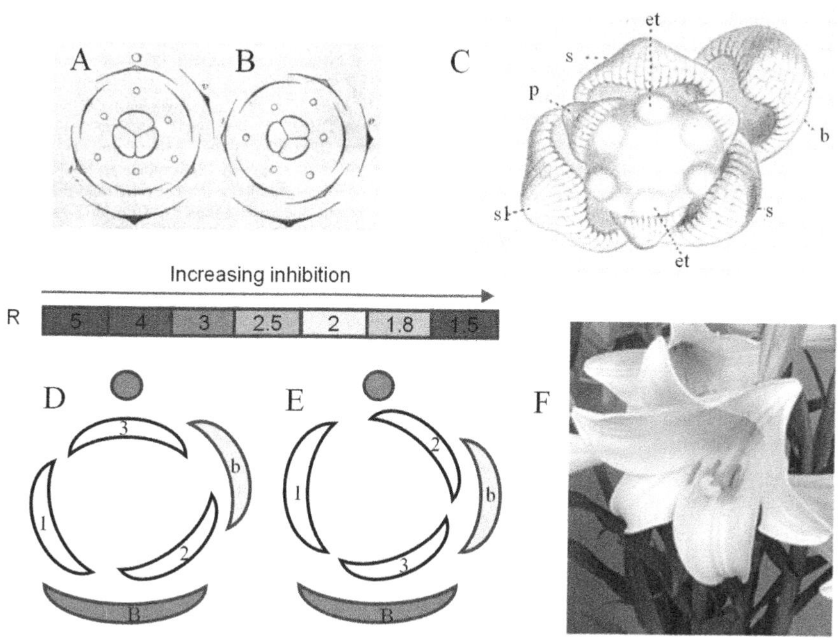

Figure 7.9. *Floral diagram of Lilium bulbiflorum: (A) with one adaxial tepal and (B) with one transverse tepal (Eichler 1875). (C) Lilium candidum (Payer 1857) with its large bracteole (b). Models with outer tepals: (D) with one adaxial tepal and (E) with one transverse tepal. The color represents the apparent distances from the sources of inhibition to the apex (R), the intensity of the inhibition being all the more important as the distance from the organ to the apex is low (see Chapter 1). (F) Mature flower of Lilium longiflorum from Andrew Dunn (2005), file under the Creative Commons Attribution license*

According to Eichler, if inflorescence phyllotaxis is spiral, K1 "falls" approximately opposite the bracteole. If the bracteole is dorso-lateral, K1 is in a diagonal ventral position and the third sepal primordium is adaxial (Figure 7.9(A)). If the bracteole is in a transverse position, K1 is transverse and opposite the bracteole (Figure 7.9(B)). The model fits the floral diagrams from Eichler and the drawing from Payer (Figures 7.9(A)–(E)).

7.5.2. *Alstroemeriaceae*

Alstroemeria

Hoshino et al. (2014) investigated *CYC*-like genes in *Alstroemeria aurea*, *A. magenta* and *A. pelegrina var. rosea*, all of which have zygomorphic flowers.

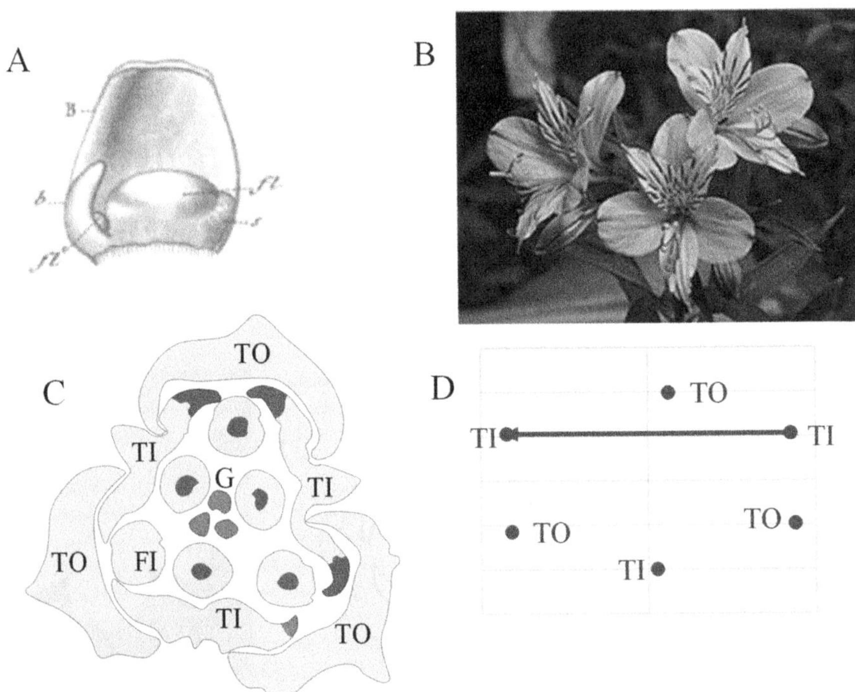

Figure 7.10. *(A) Alstroemeria versicolor according to Payer (1857), showing bracteole (b). (B) Mature flower of A. aurantiaca from JJ Harrison. File under the terms of the GNU Free Documentation License. (C) Localization of AaTCP1 transcripts detected using in situ hybridization of flower buds of A. aurea, according to Hoshino et al. (2014). TO = outer tepal; TI = inner tepal; FI = filament; G = gynoecium. (D) Model. Dark blue = expression of CYC-like gene; light blue = no expression*

To determine the temporal and spatial expression patterns of *CYC* homolog genes in *Alstroemeria*, Hoshino et al. (2014) used RT-PCR and in situ hybridization. RT-PCR indicated that *AaTCP1* is expressed in flower buds but not in leaves or stems, and that during floral development, *AaTCP1* expression is enhanced and maintained in the late stage of flower bud formation. The expression regions of *CYC* homolog genes in *Alstroemeria* were identified by hybridizing flower buds in situ using an antisense probe. Signals were observed at the adaxial region of inner perianth in accordance with Curie's principle (Figure 7.10(C)).

Organ	Angle (°)	Distance to the center
Adaxial source	90	3
B	270	4
b	150	2

Table 7.5. *Alstroemeria aurea modeling parameters*

The modeling requires forcing the constitution of twin internal tepals; otherwise, the second internal tepal would be abaxial (Figure 7.10(D)). Zygomorphy is constitutional.

7.6. Asparagales

Asparagales is a species-rich order comprising ca. 50% of all monocots, corresponding to 10%–15% of flowering plants. Orchidaceae flowers are variously bilateral and undergo extreme elaboration of some organs, including differentiation of perianth parts, stamen abortion and fusion of floral parts from the same cycle or from different cycle. In the bilateral resupinated orchid flowers, the two dorsal petals are very similar to each other; whereas, the anterior one (the lip or labellum) often undergoes extreme elaboration in shape, color, size and epidermal specializations. Such floral elaboration has been linked to extremely specialized pollination mechanisms and the exceedingly high diversification rates in Orchidaceae. Conversely, Asparagales outside orchids consists of radially symmetrical, trimerous flowers with tepaloid perianth and free floral organs (Madrigal et al. 2017).

7.6.1. *Orchidaceae*

Orchidaceae represents the most successful family of angiosperms, with species estimates ranging between 25,000 and 35,000. Despite this, the general floral Bauplan is remarkably conservative. The Orchidaceae flowers are strictly monosymmetric flowers with strong differentiation of tepals and development of an **adaxial labellum**, and various reductions of the androecium to three, two or a single stamen, with staminodes connected with the style in a monosymmetric gynostemium or column. Flowers are **resupinate** by a torsion of the pedicel at maturity, **placing the labellum in abaxial (anterior) position** and stamens and stigma in adaxial position (Ronse de Craene 2022).

According to Ronse de Craene (2022), Orchidaceae monosymmetry is "constitutional" (structural zygomorphy sensu Rudall and Bateman 2004), implying that monosymmetry is strengthened by elaboration, reduction or suppression of organs that may occur simultaneously in a same flower. Petals arise sequentially and develop a different size, while stamens may be reduced or lost.

Telipogon

The inflorescence of *T. falcatus* (Orchidaceae) is racemose, as in all orchids: in the absence of a bracteole, the source of dorsal inhibition is the apex of the inflorescence (IM shown in Figure 7.11(A)).

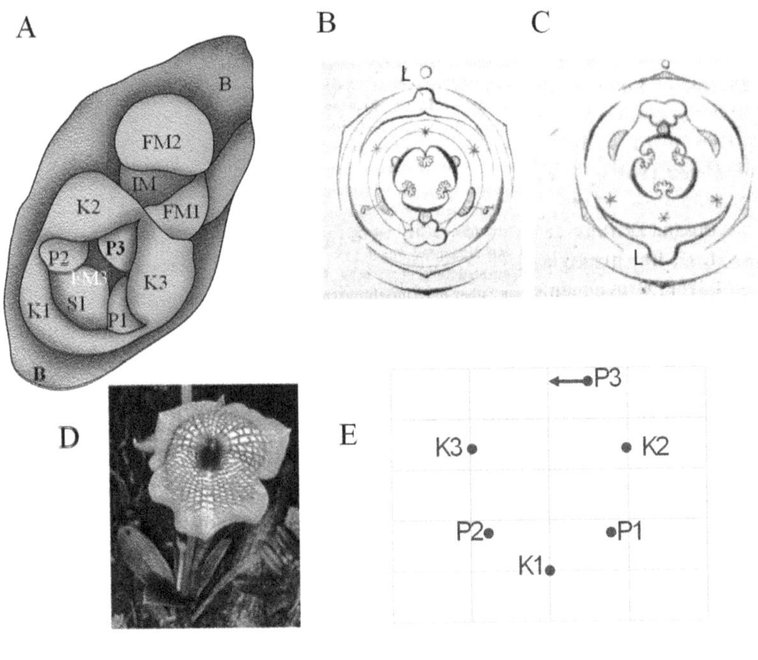

Figure 7.11. *(A) Partial inflorescence of Telipogon falcatus (from Pabón-Mora and Gonzalez 2008). IM = inflorescence meristem. FM1 = the youngest floral meristem, FM3 the oldest; P3 = labellum. (B) Floral diagram of Orchis before resupination and (C) after resupination (Eichler 1875). L = labellum. (D) Mature (resupinate) T. hausmannianus flower from Orchis, file under the Creative Commons Attribution license. (E) Model before resupination*

Unlike dicots (e.g. *Antirrhinum majus*), sepals can be initiated within very young floral meristems (FM3) because two early sources of inhibition are sufficient: the inflorescence meristem and the bract (the contact pressure of upper bracts is not necessary). The inhibition generated by the bract is weak compared to that produced by the inflorescence meristem, which results in the abaxial position of K1.

Organ	Angle (°)	Distance to the center
Inflorescence meristem	90	1.5
B	270	4

Table 7.6. *Telipogon modeling parameters*

In monocots without a bracteole, sepal primordia arrangement is the consequence of the effect of an inhibition source located on the inflorescence axis and the inhibition generated by the bract. The intensity of the adaxial source (R = 1.5) is stronger than that of the bract (R = 4). Since the two sources show bilateral symmetry, the perianth also presents bilateral symmetry (Curie's principle). The very high level of inhibition produced by the inflorescence meristem prevents the labellum (P3) from being perfectly adaxial. To respect this location, zygomorphy must be constitutional with a refocusing of the labellum (Figure 7.11(E)).

Phalaenopsis

Lin et al. (2016) have investigated by in situ hybridization the detailed spatial expression patterns of *PeCYC1* and *PeCYC2* in the *CYC/TB1* subclade in *Phalaenopsis equestris* (Orchidaceae) at the early stage of floral development.

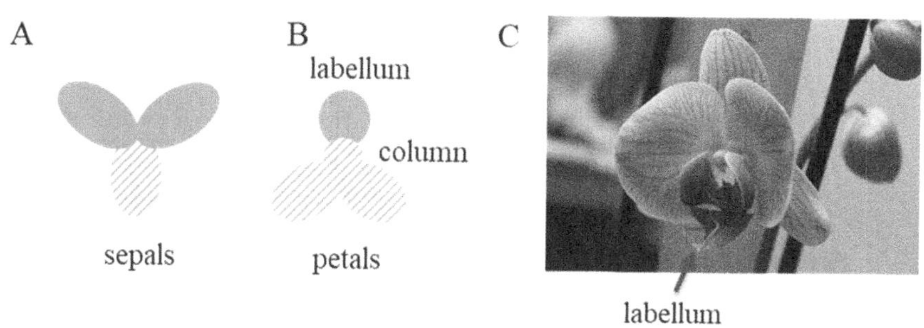

Figure 7.12. *Phalaenopsis equestris. (A and B) Diagram of PeCYC expression before resupination according to Lin et al. (2016). Blue = expression of CYC-like gene; striped = no expression. (C) Mature flower of Phalaenopsis after resupination, from Kev22, file under the Creative Commons Attribution license*

Results showed transcripts of *PeCYC1* were detected at the primordia of lateral sepal and labellum (Figure 7.12(A)). Transcripts of *PeCYC2* were concentrated at the dorsal parts of the lateral sepal and labellum primordia (before resupination). "Expression of *PeCYC*s at the floral dorsal region might correlate with orchid floral zygomorphic development" (Lin et al. 2016), which means conformity with Curie's principle.

7.6.2. *Iridaceae*

Iris

In Iridaceae, the inflorescence is a rhipidium (see Walch and Blaise 2023, Chapter 7) and the bracteole is inserted in an adaxial position and possesses two keels (Figure 7.13). The flower of *Iris* is actinomorphic but functions as three floral units through the strong development of the three styles dividing the flower in three separate sections (Ronse de Craene 2022) (Figure 7.13(C)).

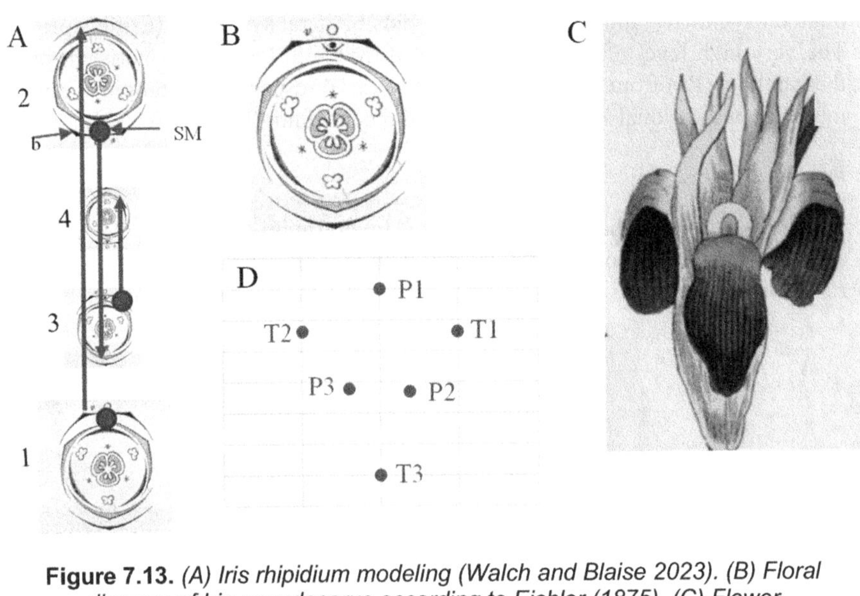

Figure 7.13. *(A) Iris rhipidium modeling (Walch and Blaise 2023). (B) Floral diagram of Iris pseudacorus according to Eichler (1875). (C) Flower of I. tuberosa from Adolphus Ypey (1813). (D) Model*

For modeling, the dorsal bikeeled bracteole is split into two close symmetrical sources of inhibition (Figure 7.13(D)).

Organ	Angle (°)	Distance to the center
b1	80	3
b2	100	3
B	270	2.5

Table 7.7. *Iris pseudacorus modeling parameters*

Gladiolus

Ronse Craene (2022) assigned positional monosymmetry to *Gladiolus* with its "subtle displacement of organs and differential growth": the orientation of stamens and style and a size difference of the petals between the adaxial and abaxial side of the flower lead this monosymmetry. Unlike to *Iris*, the inflorescence is a dichasium (Figures 7.14(A) and (B)).

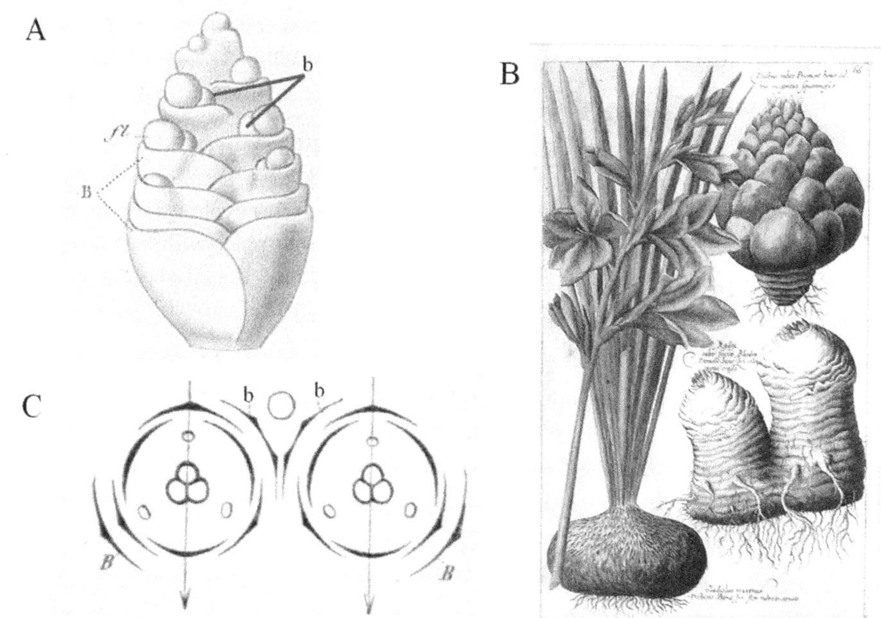

Figure 7.14. *(A) Inflorescence diagram of Gladiolus communis according to Payer (1857). B = bract; b = bracteole. (B) Gladiolus maximus from Emanuel Sweert (1612; in the public domain in the United States). (C) Partial inflorescence diagram (dichasium of two flowers) of Gladiolus cardinalis from Eichler (1875)*

The racemose inflorescence diagram of Payer (1857) shows the bracteoles converging towards the axis of the inflorescence, which we also find on the floral diagram of Eichler (1875). The model is identical to that of *Iris* (Figure 7.13(D)), but the floral axes are inclined relative to the axis of the inflorescence.

7.7. Commelinales

7.7.1. *Commelinaceae*

Commelinaceae are generally soft and fleshy herbs, some with a degree of succulence, and their inflorescences have cymose units; flowers have a differentiated perianth and are typically fugacious (short lived); their hairy anthers are well known (Soltis et al. 2018).

Tradescantia, Commelina

Preston and Hileman (2012) compared the expression of *TCP* genes in actinomorphic (*Tradescantia pallida*) and zygomorphic (*Commelina communis*) species of Commelinaceae.

Flowers arise in cincinnus. According to Eichler (1875), bracteoles of *Commelina coelestis* are Unterdrückt (Figure 7.15(A)). *Tradescantia pallida* is perfectly actinomorphic, with three identical petals and a cycle of six stamens. On the contrary, *Commelina communis* is strongly zygomorphic, with two large dorsal inner tepals (blue), reduced ventral inner tepal, reduced dorsal staminodes (upper yellow organs) and pollen-producing ventral stamens (lower yellow organs) (Figure 7.15(D)) (Preston and Hileman 2012).

Three sources of inhibition determine the arrangement of a Commelinaceae calyx (Figure 7.15(E): flower Fn): the bracteole from the former flower (b_{n-1}); the contact pressure exerted by the preceding floral meristem (F_{n-1}); the continuation sympodial meristem (SM_{n+1}), which plays the same role as the inflorescence meristem in orchids.

Preston and Hileman (2012) carried out expression analyses on dissected buds using qRT-PCR. No *TB1a* expression was detected in either outer or inner tepals of actinomorphic *Tradescantia pallida* (Figure 7.15(B)). In zygomorphic *Commelina communis* (Figure 7.15(C)), *TB1a* expression was significantly higher in ventral (K2, K3) versus dorsal outer tepals (K1), and ventral (P1) versus dorsolateral inner tepals (P2, P3). In contrast to qRT-PCR analyses, no *TB1a* expression was detectable in early to late-stage flowers of *C. communis* or *T. pallida* using in situ hybridization. This suggests that *TB1a* transcripts are below the level of detection using this method.

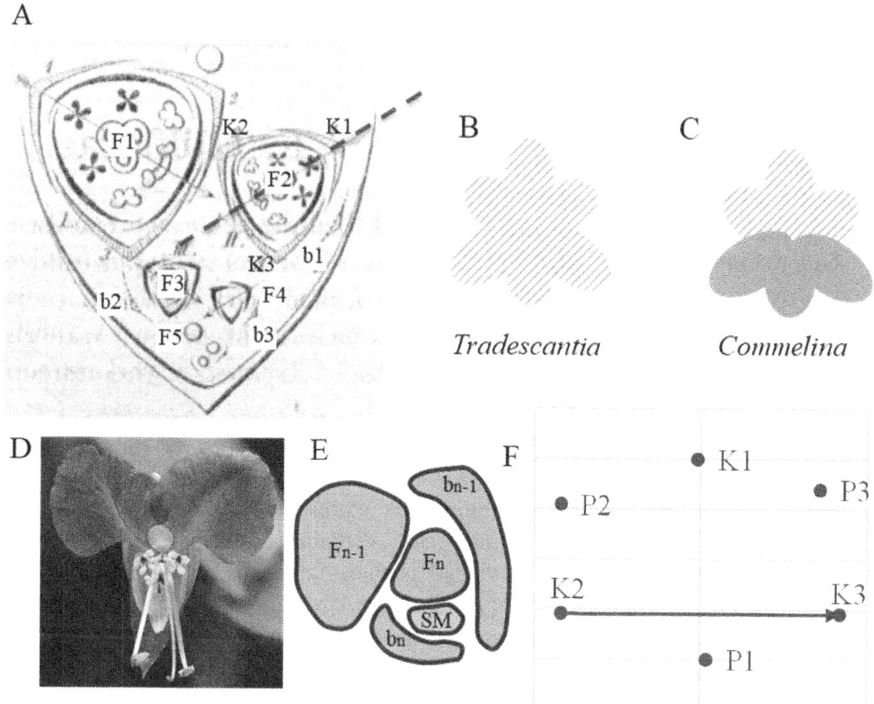

Figure 7.15. *(A) Inflorescence diagram of Commelina coelestis from Eichler (1875). b = cryptic bracteoles (dashed by Eichler). Diagram of TB1a (CYC-like) expression according to Preston and Hileman (2012) in perianth of (B) actinomorphic Tradescantia pallida and (C) zygomorphic Commelina communis. (D) C. communis mature flower from Reinhold Möller (2019), file under the Creative Commons Attribution license. (E) Initiation of floral meristem in Commelinaceae according to Hardy et al. (2004); see also Kirchoff (2003). (F) Model of Commelina perianth initiation according to Table 7.8 parameters. SM = continuation sympodial meristem initiating FM_{n+1}*

Organ	Angle (°)	Distance to the center
F_{n-1}	154	1.8
b_{n-1}	14	1.5
SM_{n+1}	260	2

Table 7.8. *Commelinaceae modeling parameters for FMn (see Figure 7.15(E); angles relative to the dorsoventral axis shown in Figure 7.15(A))*

Preston and Hileman (2012) concluded that *TB1*-like genes have been independently recruited in establishing Commelinaceae corolla bilateral symmetry in early development, similar to several core eudicot lineages and rice. However, in contrast to the asymmetric expression of *CYC* and *DICH* in the model core eudicot *A. majus* and *REP1* in *O. sativa* (rice), expression is significantly higher in the ventral, as opposed to dorsolateral, tepals. This does not change the conformity with the Curie principle, both cases showing asymmetry with respect to the transverse axis. Lack of expression of the *TB* gene in *Tradescantia pallida* corresponds to actinomorphic perianth. High ventral expression asymmetric with respect to the transverse axis in *Commelina communis* corresponds to zygomorphy.

7.8. Zingiberales

Zingiberales are with Orchidacea a monocot species-rich group with predominantly monosymmetric flowers, excluding Cannaceae and Marantaceae (Reyes et al. 2016). Perianth zygomorphy is expressed to greater and lesser degrees within the "ginger families" (the clade containing Costaceae, Zingiberaceae, Cannaceae, and Marantaceae), although in these families, the perianth is often less important in floral display than the staminode labellum. Indeed, in Costaceae, and Zingiberaceae, five androecial members develop as petaloid staminodes that fuse to form the staminodial labellum, with the fertile stamen number reduced to one adaxial outer-cycle stamen that defines the single axis of symmetry (Bartlett and Specht 2011).

7.8.1. *Cannaceae*

Canna indica

Each inflorescence of the Cannaceae consists of a main axis with spirally arranged bracts, each of which subtends a dichasium composed of a single pair of flowers (Figure 7.16(A)). In *Canna indica*, the flowers of a pair are not mirror images. The first flower gives rise to a mature flower, while the second often ceases development while it is still quite small (Figure 7.16(B)). One dorsal sepal (K1) appears first, followed by the ventral (K2) and the other dorsal (K3) sepals in that order (Kirchoff 1983).

Cannaceae is one of the few families that possesses truly **asymmetric** flowers with an asymmetric distribution and elaboration of petaloidy and organ abortion across the six members of the androecium (Figure 7.16). *Canna* flowers consist of two androecial cycles, each with three members. In the outer cycle, one androecial member develops into a mature petaloid staminode (i.e. infertile stamen); whereas, the other two primordia typically abort shortly after initiation. The inner cycle contains two petaloid staminodes plus the fertile stamen, which is in fact a "half-fertile stamen" (PA) comprising an anther with a single theca and a petaloid appendage. Of the two inner cycle staminodes, one is curved and referred to as the "labellum" (L). Both the half fertile stamen and the presence of the labellum contribute to the asymmetry of the androecium. The style is laminar, making the gynoecium appear asymmetric as well (Yu et al. 2020) (Figure 7.16(B)).

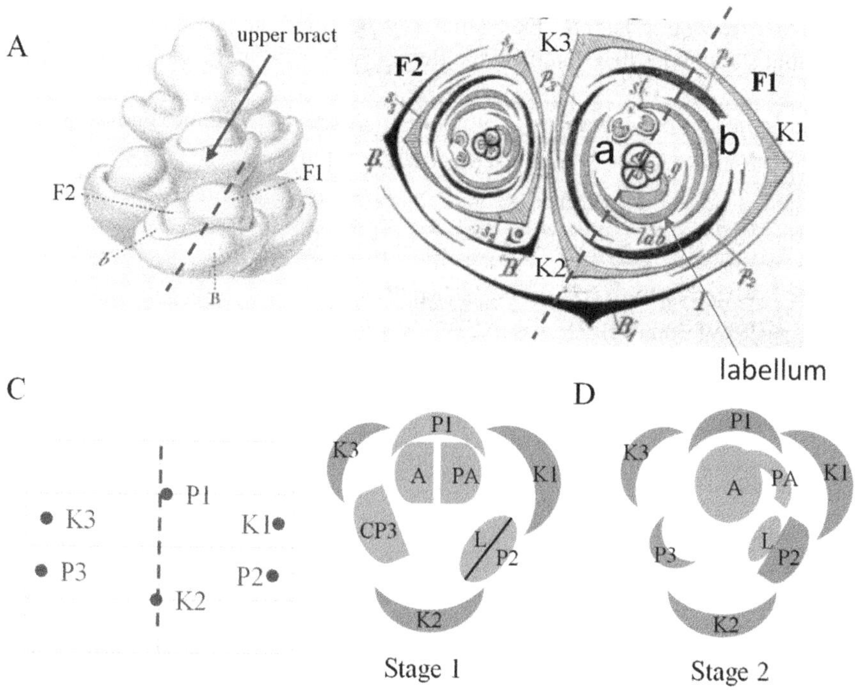

Figure 7.16. *Canna indica. (A) Inflorescence according to Payer (1857) F1 and F2 are the two flowers of the dichasium; B is the bract and b the bracteole. (B) Cyme according to Eichler (1875). a and b are the two petaloid staminodes surrounding the labellum. (C) Model of Eichler right flower (F1). (D) Expression of CiBL1A at stage 1 and stage 2 of development according to Yu et al. (2020). K = sepal; P = petal; CP = petal-stamen common primordium; A = anthere; L = labellum; PA = petaloid appendage*

As in all monocots with a lateral bracteole, the calyx is spirally symmetrical.

Yu et al. (2020) carried out RNA in situ hybridization and qRT-PCR to investigate the spatiotemporal expression patterns of gene during inflorescence and floral development in *Canna indica*. At early flower development, expression of CiBL*1A* and *CiBL1B-1* is found throughout sepals and petals. The perianth is actinomorphic at initiation (Figure 7.16(D)), which corresponds to this expression. Yu et al. (2020) found the *CYC/TB1*-like homolog *ZinTBL1B-1* exclusively in the four ginger families, which bear flowers with elaborate petaloid staminodes. They did not recover *ZinTBL1B-1* in the banana family (Musaceae), which lack a predominance of stamen-derived petaloid structures. This indicates that the presence of this particular *CYC/TB1*-like gene may be important in the formation of the petaloid staminodes that dominate the floral display within the ginger families.

Organ	Angle (°)	Distance to the center
Dorsal source (upper bract in the raceme)	120	2.5
Bract	300	2.6
Bracteole	210	2

Table 7.9. *Canna indica modeling parameters for F1. Angles relative to the dorsoventral axis of the flower (dashed red on Figure 7.16(B))*

7.8.2. *Heliconiaceae, Costaceae*

Heliconia

In *Heliconia latispatha*, the stem bears distichous orange bracts, each of which wraps a cincinnus (Figure 7.17(A)). A primary cincinnus apex is initiated in the axil of each primary bract; it forms a bracteole and terminates in a flower. A sympodial continuation meristem forms in the axil of the bracteole and repeats the same developmental pattern, as does the primary meristem (b2, F2; Figure 7.17(B)), etc. Each bracteole forms on the lateral side of the cincinnus at a 35°–40° angle to the medial axis of the cincinnus (Kirchoff 2003).

Sepals are formed in a spiral pattern; these spirals rotate clockwise in even flowers and counterclockwise in odd flowers (antidromia of uniparous cincinnus). The first sepal (K1) forms near the bracteole, the second (K2) being close to the axis of the cincinnus (the arrangement of F1 sepals in Figure 7.17(B) is the same as the model in Figure 7.15). This organization makes it possible to optimize the packing of flowers within the cyme (Kirchoff et al. 2009).

Bartlett and Specht (2011) investigated the expression patterns of *Heliconia stricta* and *Costus spicatus* orthologs of *ZinTBL1A* and *ZinTBL2* in developing flowers using RNA in situ hybridization.

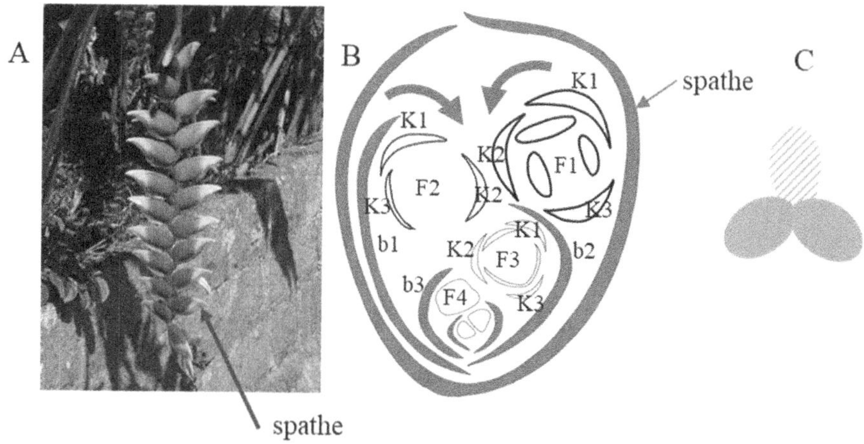

Figure 7.17. *(A) Heliconia rostrata inflorescence showing red spathose primary bracts (from Cesarious (2005), file licensed under the Creative Commons Attribution license). (B) Heliconia latispatha cincinnus (Kirchoff 2003). Arrows: even flowers are antidromous to odd flowers. b = real bracteoles. (C) HsTBL2 expression in Heliconia stricta calyx according to Bartlett and Specht (2011). Blue = expression of CYC-like genes in the floral meristem. Striped = lack of expression*

In *Heliconia stricta*, *HsTBL1A* expression was detected in the gynoecium and the posterior (dorsal) staminode. Weak expression was also detected in the petals. Expression of HsTBL*2* was found throughout the anterior (ventral) sepals and in the sepal margins (Figure 7.17(C)) (Bartlett and Specht 2011).

At maturity, *Heliconia* flowers look noticeably zygomorphic because the abaxial sepals are considerably smaller than the adaxial sepal from inception and the adaxial sepal separates from the other sepals at anthesis (Kirchoff et al. 2009). *HsTBL2* may restrict the growth of the abaxial sepals, causing zygomorphy of the sepal cycle throughout development (Bartlett and Specht 2011).

Costus

The flowers of *Costus spicatus* are strongly zygomorphic and consist of three fused sepals, a floral tube formed by the proximal fusion of the androecium and the corolla, and a trilocular inferior ovary. The petals are free distally, and the adaxial petal is larger than the lateral petals. There are two trimerous androecial cycles. A single adaxial, interior-cycle stamen is fertile. The remaining androecial members develop as petaloid staminodes and fuse to form the staminodial labellum (Bartlett and Specht 2011).

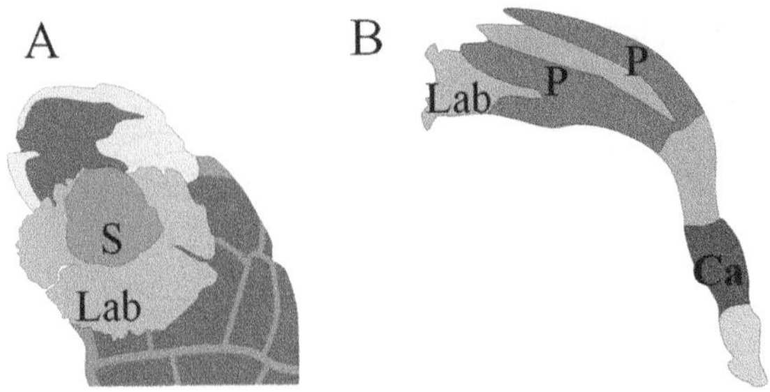

Figure 7.18. *(A and B) Costus spicatus from Bartlett and Specht (2011) showing conspicuous staminode labellum. S = fertile stamen; Ca = calyx tube; Lab = staminodial labellum; P = petal*

CsTBL1A expression was detected by Bartlett and Specht (2011) in the early floral meristem and the primary bract of *Costus spicatus*. Later in development, once floral organs were clearly discernable, *CsTBL1A* expression became restricted to the abaxial side of the flower and the region of the bract closest to the abaxial side of the flower. Expression of *CsTBL1A* in older flowers was detected in the abaxial side of the floral tube, the labellum, the gynoecium and the anther thecae. The *C. spicatus* ortholog of *ZinTBL2*, *CsTBL2*, was expressed in the fertile thecae of the single adaxial, fertile stamen. No expression of *CsTBL2* was observed in the sepals of *C. spicatus*. After examining numerous sections, Bartlett and Specht (2011) did not observe *CsTBL2* expression in the labellum, but the absence of *CsTBL2* expression in the labellum remains to be confirmed.

7.9. Poales

In grasses and sedges (Poales), the structures are often highly reduced, resulting in a partially separate descriptive terminology. In these wind-pollinated flowers, the entire inflorescence has become compressed; in some cases, it is condensed to a pseudanthium with many protruding stamens. In grasses, the influence of the **palea** on the spatial organization of the flower has resulted in a distinctive, highly reduced and compressed inflorescence morphology, which represents an obvious physical constraint to development (Rudall and Bateman 2004).

7.9.1. *Poaceae*

Poaceae (grasses) is one of the largest flowering plant families in angiosperms, with about 10,000 species and 700 genera, including many economically important crops such as rice (*Oryza*), barley (*Hordeum vulgare*) and maize (*Zea*). Evolutionary change in the organization and structure of inflorescence and flower resulted in their distinct morphology in grasses diverging from those of higher eudicots and even other monocots (Grass Phylogeny Working Group 2001). The grass inflorescence contains a number of distichous spikelets, and each spikelet has several florets subtended by a pair of glumes. Each grass floret typically consists of three types of organs: the outer bract, called the lemma; an inner bract or bracteole, called the palea; two to three lodicules and one or two cycles of three stamens (Rudall and Bateman 2004; Yuan et al. 2009).

Oryza (rice)

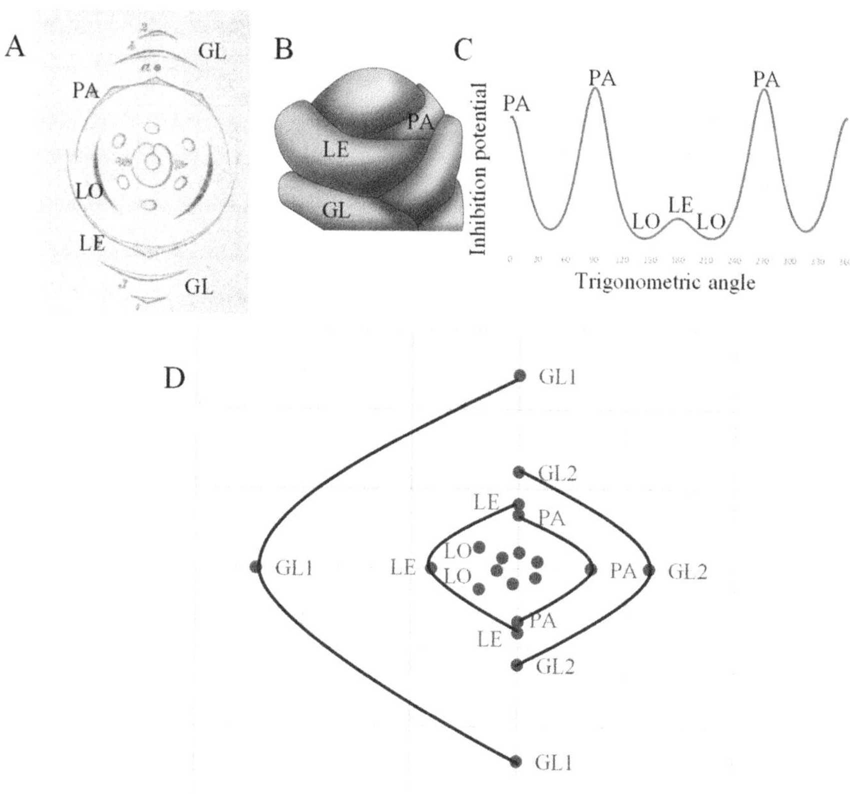

Figure 7.19. *(A) Floral diagram of Oryza sativa (Eichler 1875). (B) Rice floret meristem at stage 4 redrawn from Yuan et al. (2009). (C) Inhibition profile at lodicule initiation. (D) Model of Oryza floral meristem. GL = glumes; LE = lemma; PA = palea; LO = lodicule; central points = stamen (carpel not shown)*

Primordium name	Angle (°)	Distance/plastochron ratio	Primordium name	Plastochron ratio
GL1	180	2	PA	1
GL1	90	2	PA	1.5
GL1	270	2	LO	1
GL2	0	1	LO	1.5
GL2	90	1	S1	1
GL2	270	1.5	S2	1
LE		1	S3	1
LE		1	S4	1
LE		1.2	S5	1
PA		1	S6	1

Table 7.10. *Plastochron ratios used to model floral organ initiation in Oryza sativa*

For the modeling (Table 7.10), each glume is considered to consist of three sources of inhibition, one in its center and the other two at its ends. The second lodicule is initiated by the "dédoublement" due to the proximity of the center of the lemma. The lemma (LE) is sufficiently close to the lodicules (LO) to make its inhibition felt at the initiation of these primordia (Figure 7.19(D)). A local maximum ("hill") is created between the two maxima generated by the palea (PA), which produces two minima, where the two lodicules will be initiated (Figure 7.19(C)).

The outer cycle and the second cycle in the rice floret are not radially symmetrical; the lemma and palea form interlocking structures, with the lemma larger than the palea (Ikeda et al. 2004), and the second cycle only consists of two lodicules biased to the lemma side. As a result, the rice floret presents bilateral symmetry along the LE/PA axis (Yuan et al. 2009).

Yuan et al. (2009) showed that floral zygomorphy along the lemma-palea axis in rice (*Oryza sativa*) is partially or indirectly determined by the CYCLOIDEA (*CYC*-like) homolog *RETARDED PALEA1 (REP1)*, which regulates palea identity and development.

The earliest expression of *REP1* was observed asymmetrically in the palea primordia of the apical flowers on the primary and secondary branches at stage 4; no signal was seen in the lodicules and lemma positions at this stage. After the initiation of stamen primordia at stage 6, the *REP1* signal was clearly detected in stamens, palea and lemma, but the signal was dispersed. Around stage 8 during the development of pistil and stamen, the *REP1* signal appeared stronger, radially in stamens and vascular bundles of lemma and palea (Yuan et al. 2009).

This expression pattern shows significant divergence from other *CYC*-like genes in eudicots, which are expressed in the dorsal regions of floral meristems.

Conclusion

C.1. Phyllotaxis models

> There cannot possibly be a physical or dynamical, though there may well be a mathematical Law of Growth, which is common to, and which defines, the spiral form in the Nautilus, in the Globigerina, in the ram's horn, and in the disc of the sunflower (D'Arcy Thompson 1917).

Therefore, D'Arcy Thompson distinguishes a mathematical law of spirals that is common to the animal kingdom and the plant kingdom, to which we could add the purely material world – in particular, magnetic bodies – from physical and/or dynamic phenomena that underly this mathematical law. We have defined this mathematical law: it is a law of symmetry (Chapter 2). As for the phenomena that could be at the origin of the geometric organization in plants, we have highlighted biomolecular (auxin in the *Arabidopsis* inflorescence) and purely mechanical (pressure within racemose inflorescences) factors (Walch and Blaise 2023, Chapter 5).

Fibonacci spirals are thus a form of symmetry: they appear along *Linum* stems or in quincuncial perianths (pseudo-whorled phyllotaxis), but also along the axis of "magnetic cacti" built in a laboratory (Nisoli et al. 2009). Braun (1831) had already observed that the denser the spike, as in *Plantago major*, the more numerous the parastichies (up to mode (13, 21)). van Iterson, a precursor of modern phyllotaxis theory, found the general mathematical relationship between density and the number

For a color version of all the figures in this chapter, see www.iste.co.uk/walch/phyllotaxis2. zip.

of parastichies, but within the framework of a purely geometric model where the spike is restricted within a rigid cylinder, and where floral meristems behave like rigid spheres (van Iterson 1907). The mathematical relationship emerged from this model, but its physical origin had nothing to do with the actual plant world.

The equations that made it possible to model plant phyllotaxis were defined by Douady and Couder (1996a) and describe a repulsion or inhibition between primordia, so that the initiation of a new primordium takes place as far as possible from preexisting primordia at the periphery of the apex. This mechanism leads to results that are consistent with the empirical rule of Hofmeister (1868), according to which a new primordium arises in the largest space left available by its predecessors.

In all our models (Walch and Blaise 2023), we use the equations of Douady and Couder (1996a), but with our own specific parameterization: we take into account the distance between the apex and the primordia in the form of their "plastochron ratio" (Chapter 1). The control parameters in our model are these plastochron ratios, which are measurable quantities, as shown by Rutishauser (1998), who measured these parameters in 17 plant species. This set of parameters makes it possible to model all possible arrangements of floral organs. The equations of the model (with an exponential repulsion that decreases with distance) consider biochemical inhibition but also contact pressure between "soft" organs.

According to Rutishauser (2019), as already admitted by Darwin, phyllotaxis patterns as observable in vascular plants appear as developmental patterns that are not under the control of natural selection. There are developmental constraints ('laws of growth') that force most spiral patterns to approach the famous Fibonacci angle, which is about 137.5°. Our conclusion is significantly different: natural selection favors all "noble" phyllotaxes, but developmental constraints almost exclude those that do not converge toward the golden angle. We therefore agree with Strauss et al. (2020) who state that "the interplay of developmental constraints and potential fitness for the emergence of developmental patterns or complex traits must be considered".

As part of a "static" model, we were able to compare the packing density of stem arrangements as a function of the angle of divergence between two consecutive primordia (Walch and Blaise 2023, Chapter 3). Spiral phyllotaxis produces the densest structures, while whorled phyllotaxis forms the least dense structures (Figure C.1).

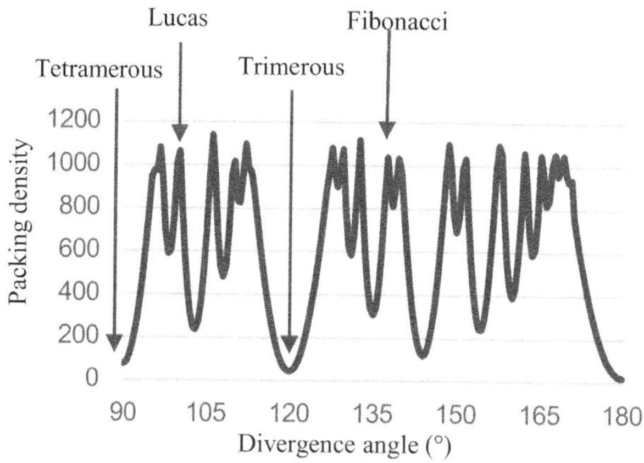

Figure C.1. *Packing density of different phyllotactic patterns (arbitrary units) in relation to the divergence angle. Lucas and Fibonacci are spiral phyllotaxis. Tetramerous and trimerous are whorled phyllotaxis*

In eudicots, and using *Linum* as an example, stem phyllotaxis begins with two opposite cotyledons (divergence angle: $\phi = 180°$). This state is stable (local minimum of potential) until the plastochron ratio reaches a critical value ($r \approx 2.2$ in blue in Figure C.2(A)) from which two minima arise, corresponding to the two opposite chiralities (clockwise and counterclockwise). Phyllotaxis becomes spiral and its mode is (1, 2). This bifurcation marks the transition from a verticillate distichous phyllotaxis to a spiral phyllotaxis. The two chiralities are equiprobable. Moving up the stem, the plastochron ratio decreases, and the number of left and right spirals increases (Figure C.2(B)) (Chapter 2).

One of the major puzzles of phyllotaxis is that Fibonacci spirals are observed much more frequently than other types of spirals, such as Lucas spirals. In angiosperms, Fibonacci spirals predominate (93.97% from a total of 8,720 observations); whereas, Lucas spirals are rare (only 1.18%) (Yin and Meicenheimer 2017).

The question then arises of why Darwinian selection seems to favor Fibonacci phyllotaxis: the answer seems to lie in the existence of developmental constraints that are independent of the genetic or environmental variations that limit organ positioning (Godin et al. 2020; Reinhart and Gola 2022).

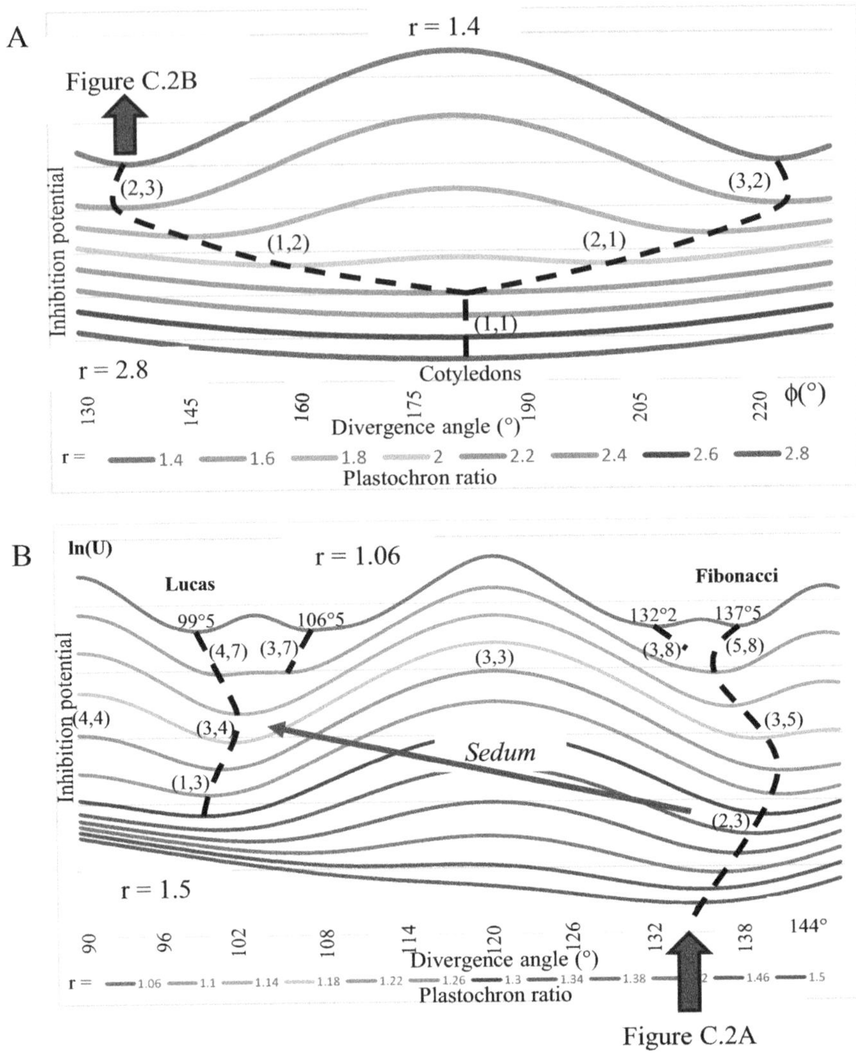

Figure C.2. *The natural log of the inhibition potential (ln(U)) as a function of the divergence angle (ϕ) for different plastochron ratios (r) calculated with the model. For each value of r, the curve of the natural log of the inhibition potential is shown; as r decreases, trajectories in phase space climb the "valleys" (local minima) formed by the curves of the inhibition potential (dashed lines). (A) Plastochron ratios ranging from 2.8 (below) to 1.4 (above) and divergence angles ranging from 130° to 225°. (B) Plastochron ratios ranging from 1.5 (below) to 1.06 (above) and divergence angles ranging from 90° to 144°. The arrow shows the transition from a Fibonacci phyllotaxis ((2, 3) mode) to a Lucas phyllotaxis ((3, 4) mode) in Sedum maximum (Chapter 2)*

Figure C.2(B) shows the dynamics of a spiral system. Only Fibonacci phyllotaxis is located in the continuity of the cotyledons. To form a system of Lucas spirals, the stem meristem must cross a potential barrier, a "mountain" located between the two "valleys" taken by the two types of phyllotaxis. We showed in Chapter 2 that this leap is equivalent to a dislocation within a crystal lattice, which is to say to the sudden interruption of a pseudo-orthostichy within a system of spirals (e.g. the stem of *Sedum*).

C.2. The bracteole theory

In 1875, the German botanist A. W. Eichler studied at length the relationship between flowers and bracteoles ("Vorblätter. Anschluss und Einsatz der Blüthe"). When only one bracteole is present – in particular, in monocots – the position of the first sepal/tepal (K1) is opposite the bracteole. When two bracteoles are present, Eichler considered that the sepals were located on the same spiral as the two bracteoles (the older bracteole α and the younger bracteole β). He then described the different possible positions of the sepals.

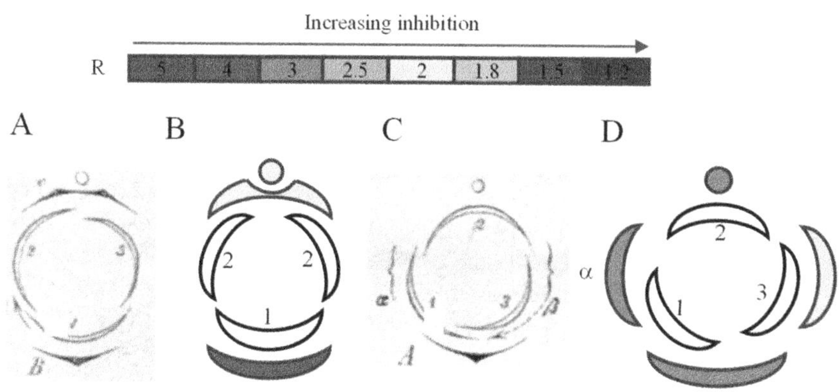

Figure C.3. *(A) Diagram of the arrangement of the outer floral organs in a monocot with a bikeeled adaxial bracteole (Eichler 1875). (B) Modeling of a flower with an abaxial bract and an adaxial bracteole (as in Iridaceae). Numbers correspond to the chronological order of tepal initiation. The upper circle represents the axis of the flower, with the adaxial bracteole exerting an influence on the position of subsequent primordia. As a result of the symmetry with respect to the dorsoventral axis, the two dorsal sepals are initiated simultaneously. (C) Diagram of a trimerous perianth in eudicots (Eichler 1875). (D) Modeling of the trimerous arrangement in a eudicot (Menispermum, Ranunculales). Colors represent the "effective distance" (R; see Box 2.1) of the inhibition sources from the center of the flower, the inhibition intensity increasing, the closer the organ producing the inhibition is to the center of the flower. Effective distances from the center of the meristem (R) are coded from violet (R = 5) to red (R = 1.2)*

If we hypothesize that bracteoles determine the arrangement of sepals, the dynamical model of phyllotaxis based on an inhibition field makes it possible to account for the numerous arrangements (trimery, tetramery and pentamery) found in monocots and eudicots. This hypothesis suggests that the first biological function of bracteoles is the determination of sepal arrangement, as intuited by Eichler in 1875. Our model of phyllotaxis explains these different cases in a "purely mechanical" way, thus responding to the challenge launched by Wilhelm Eichler 150 years ago (Walch and Blaise 2022b).

In monocots, the first sepal is initiated at 180° from the bracteole; this pattern can be considered to conform to Hofmeister's rule (Figure C.3(A)). In trimerous eudicots, however, this rule does not apply, as the distances between bracteoles α and β and the center of the floral meristem (β being closer as it is initiated after α) determine sepal arrangement (other external organs are too distant to exert an influence). In general, the greater the influence of bracteole β, the closer the first sepal is to α, that is, the more the sepal position deviates from the median position between α and β.

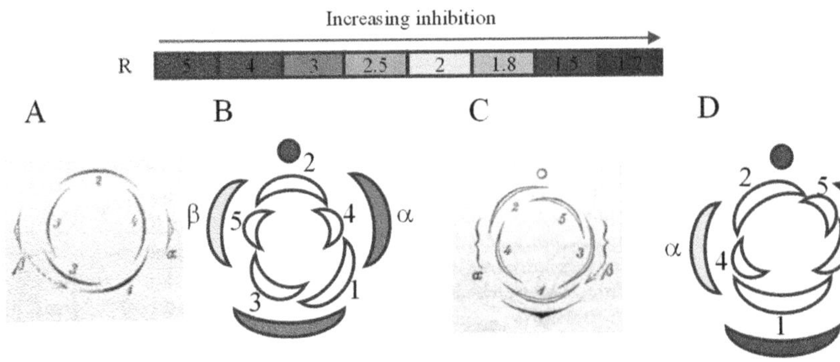

Figure C.4. *(A) Diagram of a main quincuncial sepal arrangement (Eichler 1875). (B) Modeling of a main quincuncial arrangement. Numbers correspond to the order of initiation of the sepals. (C) Diagram of the sepal arrangement found in Faboideae and Caesalpineae (Eichler 1875). (D) Modeling of the sepal arrangement found in Faboideae and Caesalpineae. Colors represent the "effective distance" (R; see Box 2.1) of the inhibition sources from the center of the flower, and the inhibition intensity increasing the closer the organ producing the inhibition is to the flower center*

In eudicots, if the bracteoles are located on the vegetative spiral (with a divergence angle close to 137°), sepals will be in a "main" quincuncial arrangement (with K2 in an adaxial position; Figure C.4(A)), which is the most frequent arrangement in eudicots. K1 is further away from bracteole β (the second bracteole in order of initiation and the closest to the center of the floral meristem), since the strength of the inhibition of β is greater than that of α (the first bracteole to be initiated and thus the furthest from the center of the floral meristem). If the strength of the inhibition of β is very high ($R_\beta = 1$; Figure C.4(D)), then K1 is abaxial, as found in Fabaceae and Caesalpinioideae.

C.3. Contact pressure mechanism

Our model takes into account not only the biomolecular inhibition potential (e.g. Figure 1.9) generated by the bracteoles in cymose inflorescences (or in certain racemose inflorescences, e.g. *Delphinium*), but also the pressure forces within the inflorescences in the absence of bracteoles.

In the racemose inflorescences of *Antirrhinum majus* and *Tropaeolum peregrinum*, it is the bracts surrounding the developing flower that exert this pressure (Figure C.5). These pressure forces are represented by Eichler (1875, 1878) by "Unterdrückt" (dashed crescent) bracteoles: an incredible intuition from this great botanist.

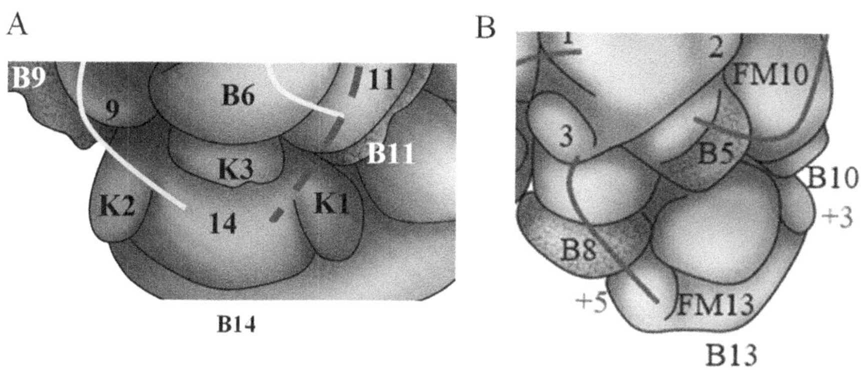

Figure C.5. *(A) The apex of the racemous inflorescence of Antirrhinum majus (redrawn from an SEM from Carpenter et al. 1995). Bracts B6, B9, B11 and B14 surround the developing floral meristem FM14. (B) Apical view of the inflorescence meristem of Tropaeolum peregrinum with the spiral sequence of initiation of floral primordia (FM numbers) in the axil of the bracts (B, followed by a number) (from Ronse de Craene and Smets 2001). The neighboring bracts that exert contact pressure on FM13 are B5 (on pseudo-orthostichy +8), B8 (on parastichy +5) and B10 (on parastichy +3)*

C.4. Curie's principle applied to phyllotaxis

According to Curie's principle, if the "causes" of phyllotaxis are symmetrical with respect to the dorsoventral axis, the arrangement of the organs that depend on them must be of at least the same symmetry. The symmetry of these causes can be of two complementary types:

1) The two bracteoles (eudicots), if any (mainly in cymose inflorescences), are transverse and generate equal inhibition potentials. Alternatively, there are no transverse sources of inhibition. Both conditions are sufficient for the flower to be symmetrical with respect to the dorsoventral axis (Figure C.6).

2) Moreover, if the inhibition potential generated by the abaxial bract and an adaxial source are of equal intensity, the flower also becomes symmetrical with respect to the transverse axis (Figure C.7).

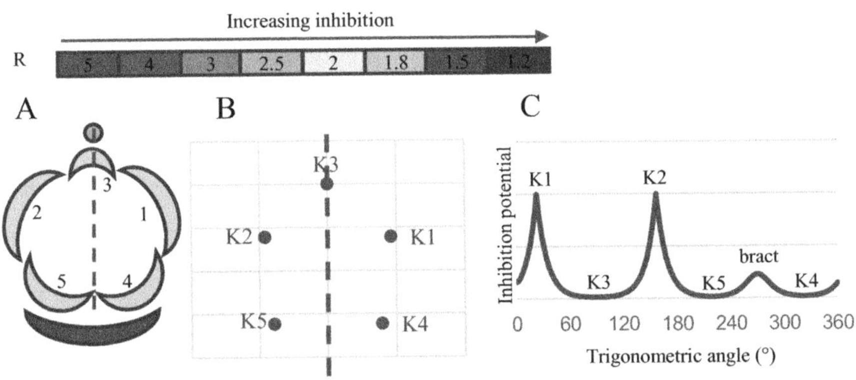

Figure C.6. *Simplified model of a pentamerous calyx without bracteoles or contact pressure. (A and B) Arrangement of sepals. (C) Inhibition profile at the initiation of K3. Dashed line = symmetry axis. Initiation of K3, K4 and K5 is simultaneous because of three equal minima of inhibition potential*

When the effective distance from the center of the floral meristem of the two axial sources of inhibition (abaxial [i.e. the bract] and adaxial) is equal to 2.5, this generates an actinomorphic tetramerous flower (Figure C.7). If the abaxial source is reinforced ($R_{Bract} = 1.3$), this produces a local maximum of the inhibition potential at $\theta = 270°$, which produces two minima in the inhibition potential curve: these give the position of the two primordia (K4, K5) on the ventral side, instead of only one ("dédoublement"; Figure C.6). The symmetry respects Curie's symmetry principle: the bilateral symmetry of the sources of inhibition is the cause of the bilateral symmetry of the calyx; the radial symmetry of the sources of inhibition is the cause of the radial symmetry of the calyx.

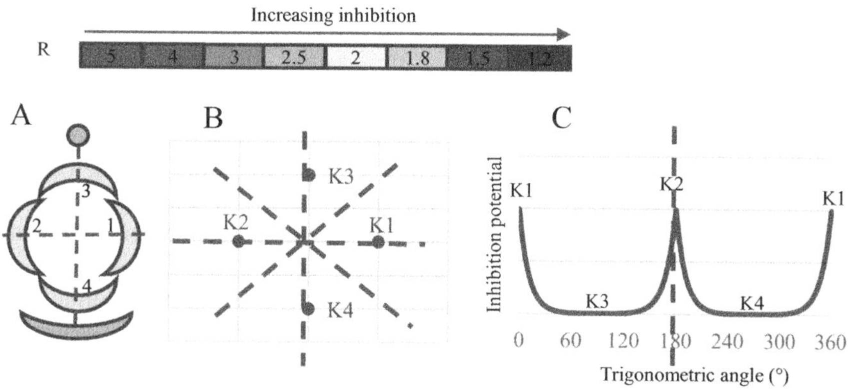

Figure C.7. *(A–C) Simplified model of a tetramerous calyx without bracteoles or contact pressure. (A and B) Diagram and model of sepal arrangement. (C) Inhibition profile at the initiation of K3. Dashed lines = axes of symmetry. Initiation of K3 and K4 is simultaneous because of two equal minima of inhibition potential*

In general, pentamery requires an asymmetry of the inhibition potential generated by the adaxial source and the bract, which makes early actinomorphy impossible; this is only possible in dimerous, tetramerous or hexamerous flowers (Figure C.8).

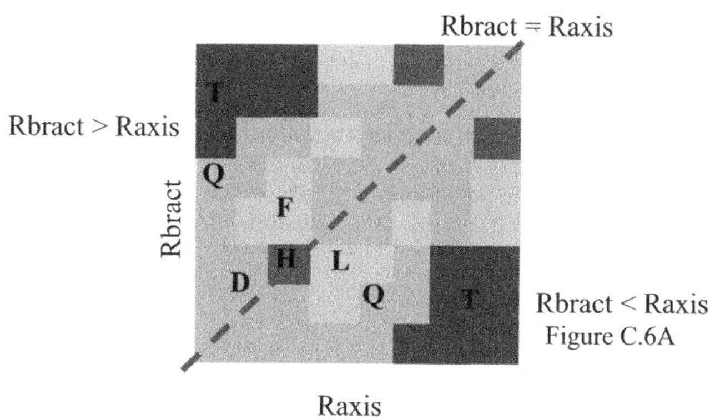

Figure C.8. *Effect of the relative intensity of adaxial and abaxial inhibition on floral organ arrangement, adapted from Nakagawa et al. (2020). D = dimerous (gray); T = trimerous (red); Q = tetramerous (green); L = pentamerous with K1 adaxial and F = pentamerous with K1 abaxial; H = hexamerous (blue)*

Naghiloo (2020) carried out a comprehensive study of floral symmetry at different developmental stages in 72 angiosperm families from 39 orders. Over the course of development, three different stages were highlighted for the expression of symmetry: organ initiation, enlargement and late differentiation. Among the 227 species examined, 56 flowers were interpreted as being actinomorphic upon organ initiation, of which 23 were trimerous, tetramerous or hexamerous. Our interpretation of the floral symmetry of most of the 33 remaining species with pentamerous flowers is that it is spiral (quincuncial) or monosymmetric at initiation, not actinomorphic. The criterion for actinomorphy applied by Naghiloo (2020) is the simultaneous appearance of petals. Our criterion for early actinomorphy is the location of the first organs at initiation (from developmental studies). The only clades included by Naghiloo (2020) that have five axes of symmetry upon floral organ initiation are Asteraceae (disc florets; Garcês et al. 2016; our Figure 5.21) and Mimosoideae (Tucker 2003): within these inflorescences, many peripheral floral meristems apply contact pressure on each developing meristem.

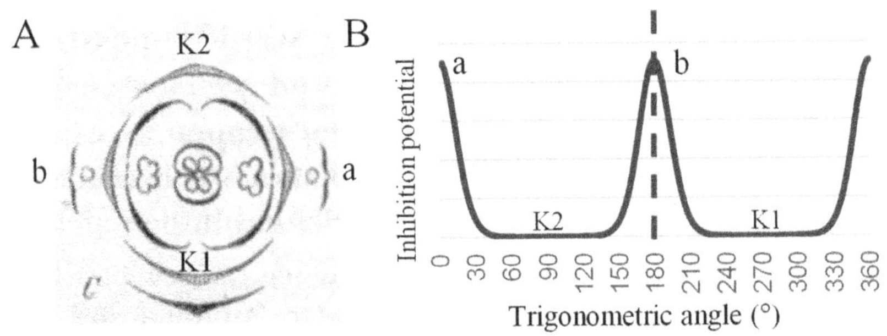

Figure C.9. *(A) Floral diagram of Olea europaea from Eichler (1875). (B) Inhibition potential profile at the initiation of the first sepals (K1, K2): the angular locations of existing maxima are those of bracteoles a (0°) and b (180°) in Figure C.9(A). Simultaneous initiation of K1 and K2 takes place at the equal potential minima (270° and 90°)*

Curie's principle applied to phyllotaxis can be stated as follows: when the symmetry of the sources of inhibition produces several minima of equal potentials, several organs are initiated simultaneously at these minima and form an arrangement of the same symmetry as that of the sources of inhibition (e.g. Figures C.6–C.9).

C.5. Equalization process

De Barros et al. (2017) noticed that in Caesalpinoids and Mimosoids, the establishment of the imbricate calyx may be related to "behavioral singularities of the developing sepals". The sepals may overlap in the intermediate stages of development if they are initiated in a spiral or bidirectionally (Ramírez-Domenech and Tucker 1990). This process is usually followed by equalization, in which sepals acquire a uniform size that allows the establishment of another type of calyx aestivation (e.g. valvate) at the end of floral development. In some rare cases, such as in *Adenanthera microsperma* (Mimosoideae; Ramírez-Domenech and Tucker 1990) and *Pentaclethra macroloba* (Mimosoideae; De Barros et al. 2017), the sepals remain imbricate. Therefore, in this latter case, an imbricate calyx can be considered to be the result of an absence of equalization in the intermediate stages of floral development and/or the maintenance of the imbricate configuration.

We modeled this process in *Illicium anisatum* (Figure 1.7) and *Ranunculus repens* (Figure 3.15). This equalization process is caused by the growth in the dimensions of organs that share a common circular space.

We showed that pentamerous perianths could not be initiated radially symmetrically, although this symmetry could be acquired during development as a consequence of the equalization process. This process is particularly visible in the change in expression of *TsCYC1C* in *Tengia* (Gesneriaceae), which is associated with a shift from zygomorphy to actinomorphy. At initiation, the perianth of *Tengia scopulorum* is irregular, its shape being elongated along the dorsoventral axis (Figure 4.3(C)). At anthesis, the perianth has acquired a quasi-circular shape (a regular pentagon = radial symmetry; Figure 4.3(E)). If *CYC*-like gene expression is asymmetric with respect to the transverse axis, as in *Antirrhinum*, the arrangement will not be able to become actinomorphic, the dorsoventral movements of the primordia being blocked. By contrast, if the functional domain of *CYC*-like genes is expanded to the entire meristem, it does not oppose to form a whorled arrangement of the primordia during development.

In *Aquilegia*, sepal initiation is clearly spiral (Figure 3.10(D)); however, sepal aestivation becomes valvate at anthesis (Figure 3.10(B)). Stamen arrangement is actinomorphic, presenting 10 perfect orthostichies. Compared to *Aquilegia*, *Delphinium* has a lower plastochron ratio between the androecium and the corolla, as well as plastochron ratios >1 between stamens that ensure the propagation of the spiral pattern from the calyx to the last stamens. In *Nigella*, the difference between the octomerous flower and the pentamerous flower is only smaller plastochron ratios for the petals and stamens (Table 3.5).

C.6. Trimery and pentamery

It has long been wondered why the main merisms in monocots and eudicots correspond to Fibonacci numbers (i.e. 3 = trimery; 5 = pentamery).

Early-diverging angiosperms either have spiral phyllotaxis or perfectly whorled phyllotaxis. The quasi-symmetrical spiral androecium of Nymphaceae is a consequence of the very low plastochron ratio between stamens (Figure 2.18).

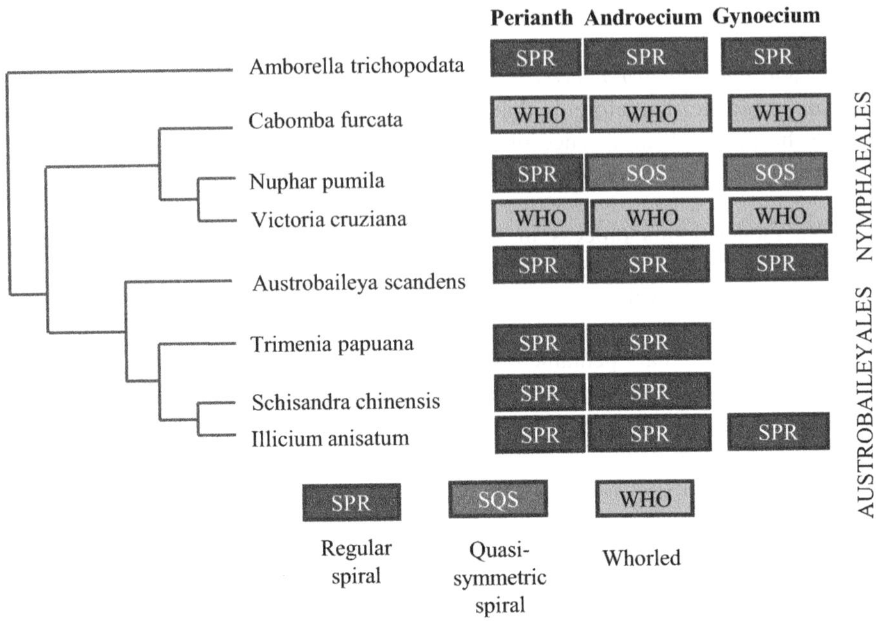

Figure C.10. *Phyllotaxis in early-diverging angiosperms from the ANA grade, according to our model. Cabomba is trimerous and Victoria tetramerous (Walch and Blaise 2022a)*

Spiral phyllotaxis optimizes the packing of primordia. There is no physical limit to the number of organs, so these numbers are labile: Austrobaileyaceae have 19–23 tepals, 7–11 stamens, 9–16 staminodes and 10–13 carpels; Trimeniaceae have up to 20 tepals, 7–16 stamens and a single carpel; *Illicium* has 7–33 tepals, 4–50 stamens and 5–21 carpels (Endress 2001). Packing is less optimal in whorled phyllotaxis, but in return, organs within a whorl are of the same type (e.g. sepal, petal, etc.); therefore, each species has a well-defined number of each type.

Magnoliales and Laurales use a “strategy” to align organ identity with spiral phyllotaxis, based on the formation of “pseudo-whorls” (termed “series” by Staedler et al. 2007), i.e. organs of the same type arranged in sets that look like whorls but are less regular because they are derived from spiral initiation with golden divergence angles (137°), instead of rational angles dividing the circle into equal parts (e.g. 72° = 360°/5). Organs within one series are not initiated simultaneously but sequentially after a short plastochron (the time interval between the initiation of two consecutive primordia), the initiation of the first organ of the next series being delayed by a longer plastochron (e.g. *Liriodendron tulipifera*, Magnoliales; Figure C.11(A)). Each cycle corresponds to the circular front of the spiral, which explains why the number of organs in a cycle is a term of the Fibonacci sequence (3, 5, 8, 13, etc.) (Figures C.11(B) and (C)). The number of organs in a series is determined by the plastochron ratio, as illustrated by our phyllotaxic frieze (Walch and Blaise 2023, Chapter 4). The androecium of *Doryphora aromatica* forms multiple series of five organs, eight in *Daphnandra repandula*, 13 in *Calycanthus floridus* and 21 in *Calycanthus occidentalis* (Laurales) (Walch and Blaise 2023).

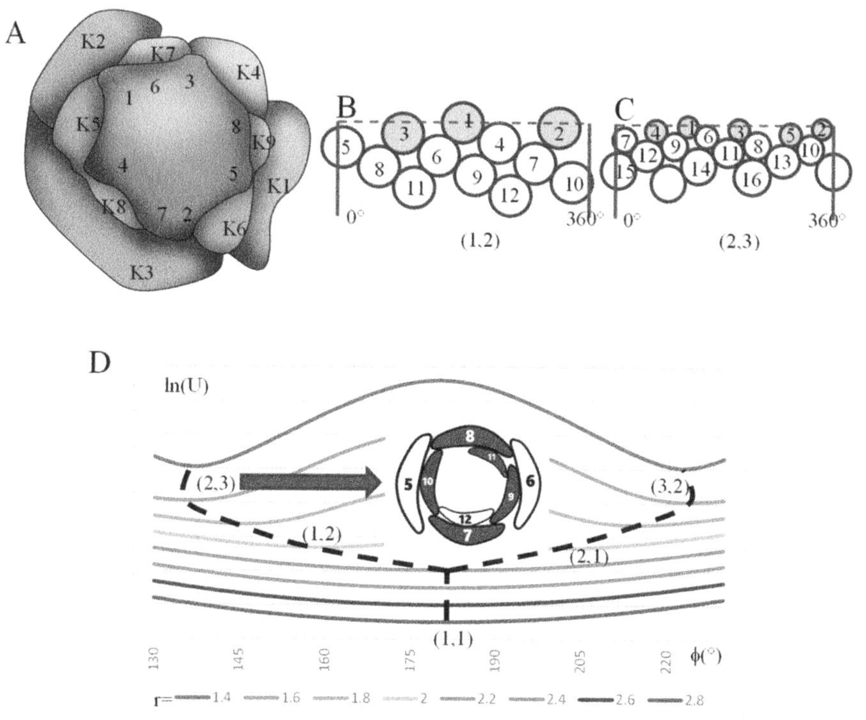

Figure C.11. *(A) The floral meristem (tepals and stamens) of Liriodendron tulipifera (Magnoliaceae), redrawn from a SEM micrograph from Erbar and Leins (1994). (B and C) Cylindrical representation of trimerous and pentamerous fronts (dashed lines) from Godin et al. (2020). (D) Pentamerous front (primordia number 7, 8, 9, 10 and 11 in black; number 12 is the next primordium to be initialized) for phyllotaxis mode (2, 3)*

For certain dimensions of the primordia compared to the meristem, spiral aestivation is of the "2/5" type (pentamery), that is, five primordia for two turns of the meristem. For larger dimensions, aestivation is of the "1/3" type (trimery), that is, three primordia for a single turn (monocots; Menispermaceae) (Figures C.11 (B) and (D)).

According to our analysis, phyllotaxis should not be coded following the classic spiral/whorled dichotomy, but according to five types: regular spiral, permuted spiral, quasi-symmetrical spiral, pseudo-whorl and whorl (Chapter 2).

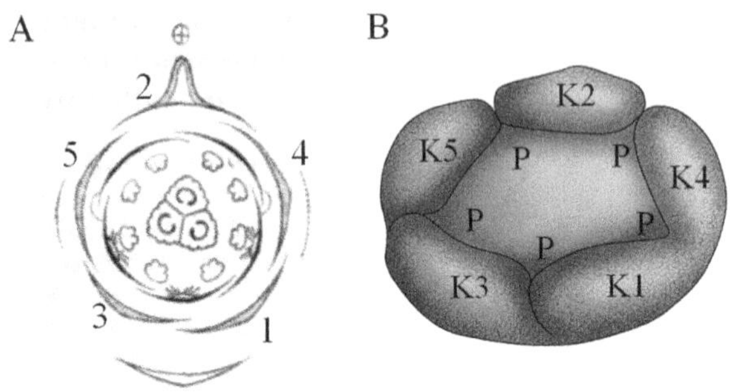

Figure C.12. *Tropaeolum majus (A) Floral diagram (Eichler 1878). (B) View of the floral bud at petal initiation, from Ronse de Craene and Smets (2001)*

For example, the quincuncial calyx of *Tropaeolum majus* is pseudo-whorled (spiral symmetry clearly visible on Eichler's floral diagram; Figure C.12).

C.7. Curie's principle applied to the genetics of floral symmetry

Curie's symmetry principle (1894), formulated in the context of electromagnetic phenomena, states that "when certain causes produce certain effects, the elements of symmetry of the causes must be found in the effects produced". We hypothesize that in angiosperms, a cause is the symmetry of the expression of *CYC*-like genes and the effect the symmetry of the corolla. There would then be four main possible cases:

– lack of expression of *CYC*-like genes: the corolla is radially symmetrical;

– expression of *CYC*-like genes throughout the corolla: the corolla is radially symmetrical;

– asymmetrical expression of *CYC*-like genes with respect to the transverse axis: the corolla is asymmetrical with respect to the transverse axis and symmetrical with respect to the dorsoventral axis due to intrinsic action of the *CYC*-like gene (zygomorphy);

– disymmetrical expression of *CYC*-like genes: the corolla is disymmetric.

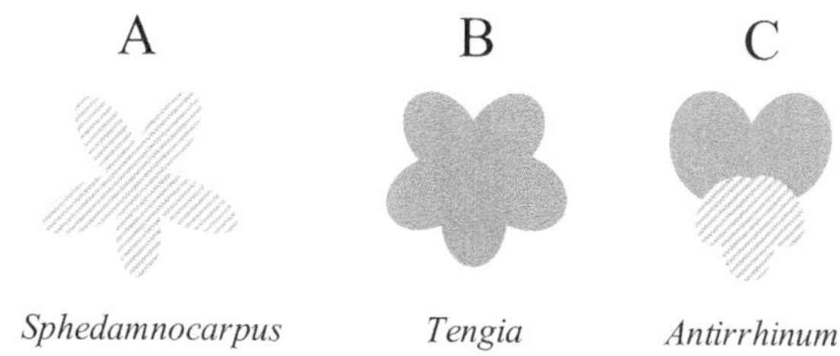

Figure C.13. *Main cause and effect relationships between the symmetry of the expression of CYC-like genes and the constitutional symmetry of the corolla: (A) lack of expression; (B) radial symmetric expression; (C) asymmetry with respect to the transverse axis*

Cause and effect relationship (expression of CYC-like genes → corolla symmetry)	**Examples observed**
Lack of expression → actinomorphy	*Tradescantia virginiana*, *Sphedamnocarpus*, *Pisum sativum* mutant, *Antirrhinum majus* peloric mutant, *Mentha longifolia*, *Senecio* disc floret
Radially symmetric expression → actinomorphy	*Canna indica*, *Cadia purpurea*, *Bergia ammannioides*, *Microsteria*, *Ligustrum lucidum*, *Tengia scopulorum*, *Plantago major*, *Veronica montana*, *Viburnum plicatum*
Asymmetry with respect to the transverse axis → zygomorphy	*Alstroemeria aurea*, *Phalaenopsis equestris*, *Commelina coelestis*, *Heliconia stricta*, *Pisum sativum*, *Stigmaphyllon*, *Tristellateia australasiae*, *Antirrhinum majus*, *Diervilla sessilifolia*, *Lonicera periclymenum*

Table C.1. *Main cause and effect relationships between the symmetry of the expression of CYC-like genes and the constitutional symmetry of the perianth (monocots and early-diverging eudicots) or the corolla (core eudicots): examples studied in this book*

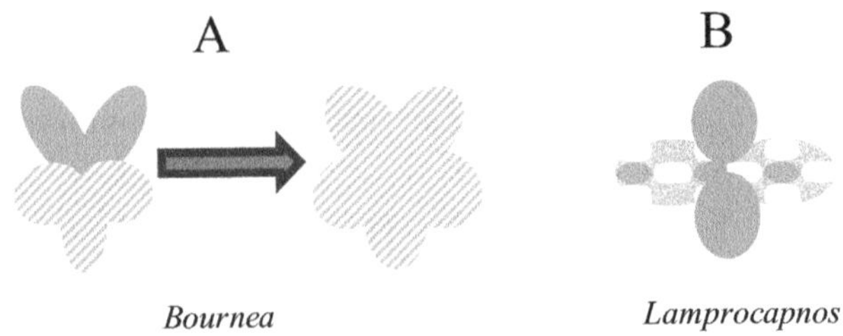

Figure C.14. *Specific cause and effect relationships between the symmetry of the expression of CYC-like genes and the symmetry of the corolla. (A) Early: dorsoventrally asymmetric expression → Late: lack of expression. (B) Disymmetric expression*

Cause and effect relationship (expression of CYC-like genes → corolla symmetry)	**Examples observed**
Early: dorsoventrally asymmetric expression → Late: lack of expression → actinomorphy	*Antirrhinum majus* backpetal mutant, *Bournea leiophylla*
Disymmetric expression → disymmetry	*Lamprocapnos spectabilis*
Late zygomorphy (elongation) (positional)	*Capnoides sempervirens, Aconitum napellus, Delphinium ajacis, Iberis amara, Tropaeolum majus, Senecio* ray florets
Late oblique zygomorphy (positional)	*Saxifraga stolonifera, Stygmaphyllon, Petunia nyctaginiflora, Salpiglossis sinuata*

Table C.2. *Specific cause and effect relationship between the symmetry of the expression of CYC-like genes and the constitutional/positional symmetry of the perianth: examples studied in this book*

Numerous examples from species belonging to different orders show that when *CYC*-like genes are expressed in the corolla, the symmetry of this expression is found in the symmetry of the corolla, in accordance with Curie's principle – if it is assumed that the expression of *CYC*-like genes is the cause, and the symmetry of the corolla, the effect (we verified this correspondence for the 84 species reviewed by Chai et al. 2023).

This conclusion is surprising, given that the *CYC* belongs to a multigenic family of transcription factors and that these genes have been recruited multiple times independently to achieve zygomorphy.

C.8. Generalized Curie's principle

In the inflorescence meristem of *Arabidopsis thaliana*, we showed (Walch 2022) that the inhibition potential corresponded to the inverse of the auxin concentration (activator), whose accumulation induces flower initiation at the periphery of the central zone of the apex (Figure 1.9).

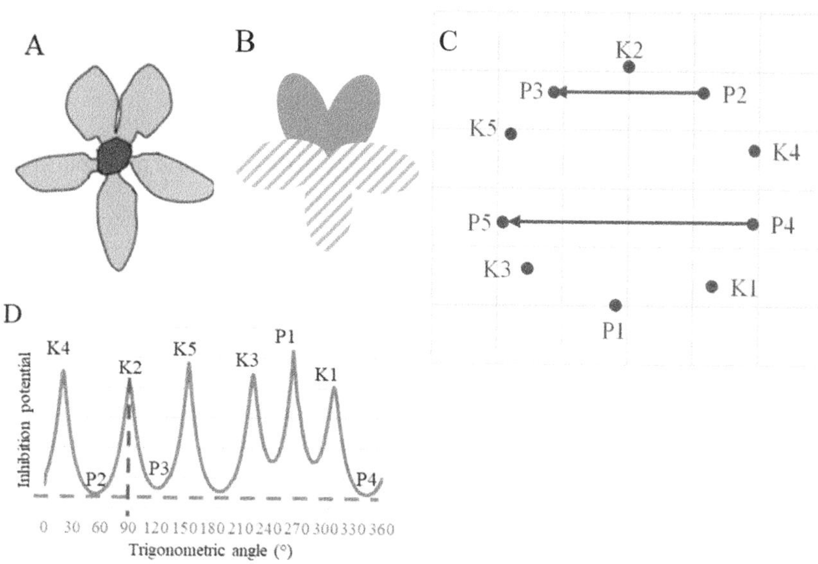

Figure C.15. *(A) Mature flower of Tristellateia australasiae. (B) CYC2A expression in the corolla of T. australasiae (Zhang et al. 2010). (C) Model. (D) Inhibition profile along the front at the initiation of P2*

Regarding the initiation of floral organs within floral meristems, things are much more complex, and we have very few experimental studies even in *Arabidopsis* (Chandler and Werr 2014; Cucinotta et al. 2021; Caumon and Vernoux 2023). Furthermore, the mechanisms at the FM interface seem diverse, since we have identified, on the one hand, the influence of bracteoles, mainly in cymes; and, on the other hand, the contact pressure exerted by the bracts surrounding each developing flower in a raceme. From a mathematical point of view, these two mechanisms can be represented by inhibitions/repulsions that decrease exponentially with the distance from the source of the inhibition or of pressure. We will simply use the general expression "phyllotaxic mechanism".

Curie's principle applies to both the phyllotaxic mechanism (section C.4) and the genetics of symmetry (section C.7). These two influences must be combined to form the perianth symmetry.

In *Tristellateia australasiae*, the calyx is quincuncial at initiation, and corolla initiation is multidirectional. At the initiation of P2, the level of the inhibition potential corresponding to the location of future P4 is lower than that of the location symmetrical to P2 with respect to the dorsoventral axis (Figure C.15(D)). For zygomorphy, in the model, we must therefore force the third petal to occupy a position symmetrical to the second petal with respect to the dorsoventral axis ("twinning": Figure C.15(C)). If not, the initiation of the petals would be totally disordered.

This rule combines the phyllotaxic mechanism and the Curie principle applied to the genetics of floral symmetry.

C.9. Constitutional versus positional zygomorphy

According to Endress (1999), there are two categories of zygomorphy:

1) "Constitutional zygomorphy" with elaborate zygomorphic flowers (e.g. Lamiales). Clades where almost all members have bilaterally symmetrical flowers have this category of zygomorphy.

2) "Positional zygomorphy" occurs in groups with predominantly actinomorphic flowers. In some cases, this zygomorphy arises by late ontogenetic modifications. According to Vöchting (1885), this monosymmetry could be induced by weight influence.

In light of these two categories of zygomorphy, Endress (1999) asks the following questions: are the shifts in floral symmetry corresponding to these different categories controlled by different mechanisms? What is the genetic difference between constitutional and positional zygomorphy? We now have the elements to answer these questions. We have identified the genetic difference between the two categories of zygomorphy: constitutional zygomorphy, with early asymmetrical expression of *CYC*-like genes, and positional zygomorphy, with late asymmetrical expression of *CYC*-like genes (Chapter 5 and Figure C.16).

C.9.1. *Constitutional zygomorphy*

Our basic model of organ initiation was successfully applied to Ranunculales (Chapter 3). However, our tests concerning zygomorphic Lamiales required a modification of the basic model. Indeed, zygomorphy is characterized by the symmetrical initiation of organs with respect to the dorsoventral axis. The necessary modification consisted of initiating two primordia simultaneously with respect to the dorsoventral axis. Initiation must be simultaneous; otherwise, the first primordium would have an influence on the position of the second. We call this process "twinning".

For constitutional zygomorphy, where monosymmetry is apparent from the onset of sepal initiation, *CYC*-like gene expression **must** be asymmetrical with respect to the transverse axis as soon as organ initiation.

C.9.2. *Positional zygomorphy*

If *CYC*-like genes are asymmetrical expressed only late in development, they can no longer modify the location of primordia, but only participate in their differential growth (Chapters 3 and 5; Figure C.16).

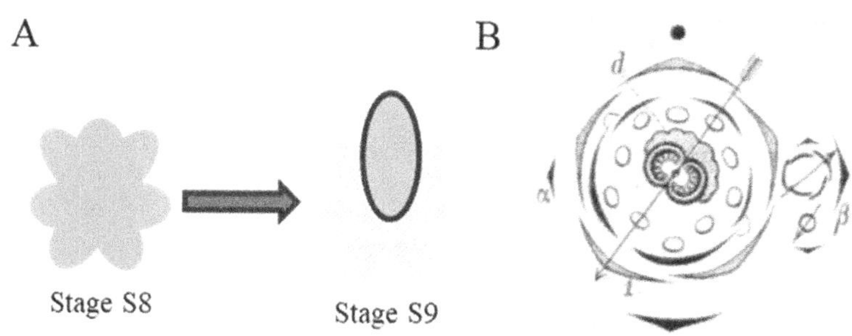

Figure C.16. *(A) Expression of DeajCYC2B and corolla symmetry in Delphinium ajacis, according to Zhao et al. (2023). Dark blue = strong expression in the spurred dorsal petal, light blue = weak expression. A late asymmetry in the expression of CYC-like genes with respect to the transverse axis appears. (B) Floral diagram of Saxifraga stolonifera showing differential growth of petal primordia along an oblique axis*

Damerval et al. (2013) showed that in Fumarioideae (Papaveraceae), asymmetric *CYC*-like gene expression is observed in the late stages of development of the flower, after zygomorphy is already visible. Therefore, unlike in constitutional zygomorphy, where the expression of *CYC*-like genes must precede the initiation of organs as it influences the position of these organs, in positional zygomorphy, *CYC*-like genes can start to be expressed after the beginning of differential growth giving rise to zygomorphy. The process is the same in *Tropaeolum* (Figures 5.16(C) and (D)).

In species with a posterior nectar spur (*Delphinium*, *Tropaeolum*, *Impatiens*), it is the balance of the flower at anthesis that is upset by this outgrowth; the spur acts as a counterweight to the rest of the flower.

In other species, positional zygomorphy is caused by the differential growth of petal primordia along an oblique axis (e.g. *Saxifraga* and *Stigmaphyllon*; see Table C.2 and Bukhari et al. 2017): the flower rotates 36°, and a single petal ends up in the posterior position. We can think that in this case, the two large ventral petals play the role of a pendulum: it is possible that the rotation during resupination takes place under the weight of these two petals.

We have too few genetic studies concerning positional zygomorphy (less than 10), compared to around a hundred for constitutional zygomorphy. Additional analyses should therefore be conducted to clarify the genetic causes of positional zygomorphy. Although positional zygomorphy is observed in orders with predominantly actinomorphic flowers, constitutional zygomorphy is found not only in predominantly zygomorphic orders (monocots [e.g. Asparagales], Fabales and Lamiales), but also in more diverse orders, such as Malphigiales (*Tristellateia*) and Dipsacales (*Lonicera*).

References

Adler, I., Barabé D., Jean R. (1997). A history of the study of phyllotaxis. *Annals of Botany*, 80, 231–244.

Almeida, J., Rocheta, M., Galego, L. (1997). Genetic control of flower shape in Antirrhinum majus. *Development*, 124, 1387–1392.

Ampornpan, L.A. and Armstrong J.E. (2002). Floral ontogeny of Salpiglossis (Solanaceae) and the oblique gynoecium. *Journal of the Torrey Botanical Society*, 129, 85–95.

APG IV (2016). An update of the angiosperm phylogeny group classification for the orders and families of flowering plants: APG IV. *Botanical Journal of the Linnean Society*, 81, 1–20.

Barboza, G.E., Hunziker, A.T., Bernardello, G., Cocucci, A.A., Moscone, A.E., Carrizo Garćia, C., Fuentes, V. (2016). Solanaceae. In *Flowering Plants. Eudicots*, Kadereit, J.W. and Bittrich, V. (eds). Springer, Cham.

Bartlett, M.E. and Specht, C. (2011). Changes in expression pattern of the teosinte branched1-like genes in the Zingiberales provide a mechanism for evolutionary shifts in symmetry across the order. *American Journal of Botany*, 98, 227–243.

Bello, M.A., Rudall, P.J., González, F., Fernández-Alonso, J.L. (2004). Floral morphology and development in Aragoa (Plantaginaceae) and related members of the order Lamiales. *International Journal of Plant Sciences*, 165, 723–738.

Bello, M.A., Hawkins, J.A., Rudall, P.J. (2007). Floral morphology and development in Quillajaceae and Surianaceae (Fabales), the species-poor relatives of Leguminosae and Polygalaceae. *Annals of Botany*, 100, 1491–1505.

Bello, M.A., Hawkins, J.A., Rudall, P.J. (2010). Floral ontogeny in Polygalaceae and its bearing on the homologies of keeled flowers in Fabales. *International Journal of Plant Sciences*, 171, 482–498.

Braun, A. (1831). *Vergleichende Untersuchung über die Ordnung der Schuppen an den Tannenzapfen als Einleitung zur Untersuchung der Blattstellung überhaupt*. Thormann, Bonn.

Bravais, A. (1850). Mémoire sur les systèmes formés par les points distribués régulièrement sur un plan ou dans l'espace. *Journal de l'école polytechniquie*, 19, 1–128.

Bravais, L. and Bravais, A. (1837). Essai sur la disposition symétrique des inflorescences. *Annales des sciences naturelles (botanique)*, 7, 193–221, 291–348.

Bruneau, A., Klitgaard, B.B., Prenner, G., Fougère-Danezan, M., Tucker, S.C. (2014). Floral evolution in the Detarieae (Leguminosae): Phylogenetic evidence for labile floral development in an early-diverging legume lineage. *International Journal of Plant Sciences*, 175, 392–417.

Bukhari, G., Zhang, J., Stevens, P.F., Zhang, W. (2017). Evolution of the process underlying floral zygomorphy development in pentapetalous angiosperms. *American Journal of Botany*, 104, 1846–1856.

Bull-Hereñu, K. and Claßen-Bockhoff, R. (2011). Open and closed inflorescences: More than simple opposites. *Journal of Experimental Botany*, 62, 79–88.

Bull-Hereñu, K., Dos Santos, P., Toni, J.F.G., El Ottra, J.H.L., Thaowetsuwan, P., Jeiter, J. Ronse De Craene L.P., Iwamoto, A. (2022). Mechanical forces in floral development. *Plants*, 11(5), 661.

Busch, A. and Zachgo, S. (2007). Control of corolla monosymmetry in the Brassicacea Iberis amara. *The Proceedings of the National Academy of Sciences*, 104, 16714–16719.

Busch, A. and Zachgo, S. (2009). Flower symmetry evolution: Towards understanding the abominable mystery of angiosperm radiation. *Bioessays*, 31, 1181–1190.

Buzgo, M. and Endress, P.K. (2000). Floral structure and development of acoraceae and its systematic relationships with basal angiosperms. *International Journal of Plant Sciences*, 161, 23–41.

Caris, P.L., Geuten, K.P., Janssens, S.B., Smets, E.F. (2006). Floral development in three species of impatiens (Balsaminaceae). *American Journal of Botany*, 93, 1–14.

Carpenter, R. and Coen, E. (1990). Floral homeotic mutations produced by transposon mutagenesis in Antirrhinum majus. *Genes and Development*, 4, 1483–1493.

Carpenter, R., Copsey, L., Vincent, C., Doyle, S., Magrath, R., Coen, E. (1995). Control of flower development and phyllotaxy by meristem identity genes in Antirrhinum. *The Plant Cell*, 7, 2001–2011.

Carrive, L., Domenech, B., Sauquet, H., Jabbour, F., Damerval C., Nadot, S. (2020). Insights into the ancestral flowers of Ranunculales. *Botanical Journal of the Linnean Society*, 194, 23–46.

Caumon, H. and Vernoux, T. (2003). A matter of time: Auxin signaling dynamics and the regulation of auxin responses during plant development. *Journal of Experimental Botany*, 74, 3887–3902.

Chai, Y., Liu, H., Chen, W., Guo, C., Chen, H., Cheng, X., Chen, D., Luo, C., Zhou, X., Huang, C. (2023). Regulation of floral development by CYC-like genes. *Current Issues in Molecular Biology*, 45, 2035–2059.

Chandler, J.W. and Werr, W. (2014). Arabidopsis floral phytomer development: Auxin response relative to biphasic modes of organ initiation. *Journal of Experimental Botany*, 65, 3097–3110.

Chang, H., Downie, S., Peng, H., Sun, F. (2019). Floral organogenesis in three members of the tribe Delphinieae (Ranunculaceae). *Plants*, 8, 493.

Chapman, M.A., Leebens-Mack, J., Burke, J.M. (2008). Positive selection and expression divergence following gene duplication in the sunflower CYCLOIDEA gene family. *Molecular Biology and Evolution*, 25, 1260–1273.

Choob, V. (2022). Prophyll in monocots: The starting point of lateral shoot phyllotaxis. *Frontiers in Plant Science*, 13, 855146.

Church, A.H. (1904). *On the Relation of Phyllotaxis to Mechanical Laws*. Williams and Norgate, London.

Citerne, H.L., Möller, M., Cronk, Q.C.B. (2000). Diversity of cycloidea-like genes in Gesneriaceae in relation to floral symmetry. *Annals of Botany*, 86, 167–176.

Citerne, H.L., Pennington, R.T., Cronk, Q.C.B. (2006). An apparent reversal in floral symmetry in the legume Cadia is a homeotic transformation. *The Proceedings of the National Academy of Sciences*, 103, 12017–12020.

Citerne, H., Jabbour, F., Nadot, S., Damerval, C. (2010). The evolution of floral symmetry. *Advances in Botanical Research*, 54, 85–137.

Citerne, H., Reyes, E., Le Guilloux, M., Delannoy, E., Simonnet, F., Sauquet, H., Weston, P.H., Nadot, S., Damerval, C. (2017). Characterization of CYCLOIDEA-like genes in Proteaceae, a basal eudicot family with multiple shifts in floral symmetry. *Annals of Botany*, 119, 367–378.

Clark, J.I. and Coen, E.S. (2002). The cycloidea gene can respond to a common dorsoventral prepattern in Antirrhinum. *The Plant Journal*, 30, 639–648.

Claßen-Bockhoff, R. and Bull-Hereñu, K. (2013). Towards an ontogenetic understanding of inflorescence diversity. *Annals of Botany*, 112, 1523–1542.

Corley, S.B., Carpenter, R., Copsey, L., Coen, E. (2005). Floral asymmetry involves an interplay between TCP and MYB transcription factors in Antirrhinum. *The Proceedings of the National Academy of Sciences*, 102, 5068–5073.

Couder, Y. (1998). Initial transitions, order and disorder in phyllotactic patterns: The ontogeny of *Helianthus annuus*. A case study. *Acta Societatis Botanicorum Poloniae*, 67, 129–150.

Cubas, P., Lauter, N., Doebley, J., Coen, E. (1999). The TCP domain: A motif found in proteins regulating plant growth and development. *The Plant Journal*, 18, 215–222.

Cubas, P., Coen, E., Zapater, J.M. (2001). Ancient asymmetries in the evolution of flowers. *Current Biology*, 11, 1050–1052.

Cucinotta, M., Cavalleri, A., Chandler, J.W., Colombo, L. (2021). Auxin and flower development: A blossoming field. *Cold Spring Harbor Perspectives in Biology*, 13, a039974.

Cui, M.L., Copsey, L., Green, A.A., Bangham, J.A., Coen, E. (2010). Quantitative control of organ shape by combinatorial gene activity. *PLOS Biology*, 8, e1000538.

Cunnell, G. (1958). Aestivation in *Ranunculus repens* L. *New Physiologist*, 57, 340–352.

Cunnell, G. (1960). Further observations on aestivation in *Ranunculus repens* L. *New Physiologist*, 63, 1–11.

Curie, P. (1894). Sur la symétrie dans les phénomènes physiques, symétrie d'un champ électrique et d'un champ magnétique. *Journal de physique théorique et appliquée*, 3, 393–417.

Damerval, C. and Nadot, S. (2007). Evolution of perianth and stamen characteristics with respect to floral symmetry in Ranunculales. *Annals of Botany*, 100, 631–640.

Damerval, C., Le Guilloux, M., Jager, M., Charon, C. (2007). Diversity and evolution of CYCLOIDEA-Like TCP genes in relation to flower development in Papaveraceae. *Plant Physiology*, 143, 759–772.

Damerval, C., Citerne, H., Le Guilloux, M., Domenichini, S., Dutheil, J., Ronse de Craene, L., Nadot, S. (2013). Asymmetric morphogenetic cues along the transverse plane: Shift from disymmetry to zygomorphy in the flower of Fumarioideae. *American Journal of Botany*, 100, 391–402.

D'Arcy Thompson, W. (1917). *On Growth and Form*. Cambridge University Press, Cambridge.

Davis, C.C. and Chase, M.W. (2004). Elatinaceae are sister to Malpighiaceae; Peridiscaceae belong to Saxifragales. *American Journal of Botany*, 91, 262–273.

Davis, C.C., Bell, C.D., Mathews, S., Donoghue, M.J. (2002). Laurasian migration explains Gondwanan disjunctions: Evidence from Malpighiaceae. *The Proceedings of the National Academy of Sciences*, 99, 6833–6837.

De Barros, T.C., Pedersoli, G.D., Paulino, J.V., Teixeira, S.P. (2017). In the interface of caesalpinioids and mimosoids: Comparative floral development elucidates shared characters in Dimorphandra mollis and Pentaclethra macroloba (Leguminosae). *American Journal of Botany*, 104, 218–232.

Deschatres, R. (1954). Recherches sur la phyllotaxie du genre Sedum. *Revue générale de botanique*, 61, 501–570.

Dickson, A. (1871). On the phyllotaxis of Lepidodendron and the allied, if not identical, genus Knorria. *Journal of Botany, British and Foreign*, 9, 166–167.

Doebley, J., Stec, A., Hubbard, L. (1997). The evolution of apical dominance in maize. *Nature*, 386, 485–488.

Dong, Z., Zhao, Z., Liu, C., Luo, J., Yang, J., Huang, W., Hu, X., Wang, T.L., Luo, D. (2005). Floral patterning in *Lotus japonicus*. *Plant Physiology*, 137, 1272–1282.

Douady, S. and Couder, Y. (1992). Phyllotaxis as a physical self-organized growth process. *Physical Review Letters*, 68, 2098–2101.

Douady, S. and Couder, Y. (1996a). Phyllotaxis as a dynamical self organizing process. Part I: The spiral modes resulting from time-periodic iterations. *Journal of Theoretical Biology*, 178, 255–274.

Douady, S. and Couder, Y. (1996b). Phyllotaxis as a dynamical self-organizing process. Part II: The spontaneous formation of a periodicity and the coexistence of spiral and whorled patterns. *Journal of Theoretical Biology*, 178, 275–294.

Douady, S. and Golé, C. (2016). Fibonacci or quasi-symmetric phyllotaxis. Part II: Botanical observations. *Acta Societatis Botanicorum Poloniae*, 85, 3534.

Eichler, A.W. (1875). *Blüthendiagramme construirt und erläutert*. Volume I. Engelmann, Leipzig.

Eichler, A.W. (1878). *Blüthendiagramme construirt und erläutert*. Volume II. Engelmann, Leipzig.

El, E., Remizowa, M.V., Sokoloff, D.D. (2020). Developmental flower and rhizome morphology in Nuphar (Nymphaeales): An interplay of chaos and stability. *Frontiers in Cell and Developmental Biology*, 8, 303.

Endress, P.K. (1998). Antirrhinum and Asteridae evolutionary changes of floral symmetry. *Symposium of the Society of Experimental Biolology*, 51, 133–140.

Endress, P.K. (1999). Symmetry in flowers: Diversity and evolution. *International Journal of Plant Sciences*, 160, S3–S23.

Endress, P.K. (2001). The flowers in extant basal angiosperms and inferences on ancestral flowers. *International Journal of Plant Sciences*, 162, 1111–1140.

Endress, P.K. (2012). The immense diversity of floral monosymmetry and asymmetry across angiosperms. *Botanical Review*, 78, 345–397.

Endress, P.K. and Armstrong, J. (2011). Floral development and floral phyllotaxis in Anaxagorea (Annonaceae). *Annals of Botany*, 108, 835–845.

Endress, P.K. and Doyle, J. (2015). Ancestral traits and specializations in the flowers of the basal grade of living angiosperms. *Taxon*, 64, 1093–1116.

Erbar, C. and Leins, P. (1994). Flowers in Magnoliidae and the origin of flowers in other subclasses of the angiosperms. Part I: The relationships between flowers of Magnoliidae and Alismatidae. *Plant Systematics and Evolution*, S8, 193–208.

Flore Virtuelle d'Orsay (n.d.). [Online]. Available at: https://www.floredorsay.universite-paris-saclay.fr/flore_web.

Galván-Ampudia, C., Cerutti, G., Legrand, J., Brunoud, G., Martin-Arevalillo, R., Azais, R., Bayle, V., Moussu, S., Wenzl, C., Vernoux, T. et al. (2020). Temporal integration of auxin information for the regulation of patterning. *eLife*, 9, e55832.

Garcês, H.M.P., Spencer, V.M., Kim, M. (2016). Control of floret symmetry by RAY3, SvDIV1B, and SvRAD in the capitulum of *Senecio vulgaris*. *Plant Physiology*, 171, 2055–2068.

Godin, C., Golé, C., Douady, S. (2020). Phyllotaxis as geometric canalization during plant development. *Development*, 147, dev165878.

Golé, C., Dumais, J., Douady, S. (2016). Fibonacci or quasi-symmetric phyllotaxis. Part I: Why? *Acta Societatis Botanicorum Poloniae*, 85, 3533.

Grass Phylogeny Working Group (2001). Phylogeny and subfamilial classification of the grasses (Poaceae). *Annals of the Missouri Botanical Garden*, 88, 373–457.

Hardy, C.R., Davis, J.I., Stevenson, D.W. (2004). Floral organogenesis in Plowmanianthus (Commelinaceae). *International Journal of Plant Sciences*, 165, 511–519.

Harris, E.M. (1995). Inflorescence and floral ontogeny in Asteraceae: A synthesis of historical and current concepts. *Botanical Review*, 61, 93–278.

Hellwig, H., Engelmann, R., Deussen, O. (2006). Contact pressure models for spiral phyllotaxis and their computer simulation. *Journal of Theoretical Biology*, 240, 489–500.

Hileman, L.C. (2014). Trends in flower symmetry evolution revealed through phylogenetic and developmental genetic advances. *Philosophical Transactions of the Royal Society London B*, 369, 20130348.

Hodges, S.A. and Arnold, M.L. (1995). Spurring plant diversification: Are floral nectar spurs a key innovation? *Proceedings: Biological Sciences*, 262, 343–348.

Hofmeister, W. (1868). *Allgemeine Morphologie der Gewachse*. Engelmann, Leipzig.

Horn, S., Pabon-Mora, N., Theuß, V.S., Busch, A., Zachgo, S. (2015). Analysis of the CYC/TB1 class of TCP transcription factors in basal angiosperms and magnoliids. *The Plant Journal*, 81, 559–571.

Hoshino, Y., Igarashi, T., Ohshima, M., Shinoda, K., Murata, N., Kanno, A., Nakano, M. (2014). Characterization of CYCLOIDEA-like genes in controlling floral zygomorphy in the monocotyledon Alstroemeria. *Scientia Horticulturae*, 169, 6–13.

Howarth, D.G. and Donoghue, M.J. (2006). Phylogenetic analysis of the "ECE" (CYC/TB1) clade reveals duplications predating the core eudicots. *The Proceedings of the National Academy of Sciences*, 103, 9101–9106.

Howarth, D.G., Martins, T., Chimney, E. (2011). Diversification of CYCLOIDEA expression in the evolution of bilateral flower symmetry in Caprifoliaceae and Lonicera (Dipsacales). *Annals of Botany*, 107, 1521–1532.

Ikeda, K., Nagasawa, N., Nagato, Y. (2004). Developmental course of inflorescence and spikelet in rice. *Breeding Science*, 54, 147–156.

van Iterson, G. (1907). *Mathematische und Mikroskopisch-Anatomische Studien über Blattstellungen nebst Betraschtungen über den Schalenbau der Miliolinen*. Gustav Fischer, Jena.

Jabbour, F., Ronse De Craene, L.P., Nadot, S., Damerval, C. (2009). Establishment of zygomorphy on an ontogenic spiral and evolution of perianth in the tribe Delphinieae (Ranunculaceae). *Annals of Botany*, 104, 809–822.

Jabbour, F., Cossard, G., Le Guilloux, M., Sannier, J., Nadot, S., Damerval, C. (2014). Specific duplication and dorsoventrally asymmetric expression patterns of Cycloidea-like genes in zygomorphic species of Ranunculaceae. *PLOS ONE*, 9, e95727.

Jean, R.V. (1994). *Phyllotaxis. A Systemic Study in Plant Morphogenesis*. Cambridge University Press, Cambridge.

Kampny, C.M., Dickinson, T.A., Dengler, N.G. (1993). Quantitative comparison of floral development in Veronica chamaedrys and Veronicastrum virginicum (Scrophulariaceae). *American Journal of Botany*, 80, 449–460.

Kim, M., Cui, M-L., Cubas, P., Gillies, A., Lee, K., Chapman, M.A., Abbott, R.J., Coen, E. (2008). Regulatory genes control a key morphological and ecological trait transferred between species. *Science*, 322, 1116–1119.

Kirchoff, B. (1983). Floral organogenesis in five genera of the marantaceae and in Canna (Cannaceae). *American Journal of Botany*, 70, 508–523.

Kirchoff, B. (1988). Inflorescence and flower development in Costus scaber (Costaceae). *Canadian Journal of Botany*, 62, 339–345.

Kirchoff, B. (2003). Shape matters: Hofmeister's rule, primordium shape, and flower orientation. *International Journal of Plant Sciences*, 164, 505–517.

Kirchoff, B.K., Lagomarsino, L.P., Newman, W.H., Bartlett, M.E., Specht, C.D. (2009). Early floral development of Heliconia latispatha (Heliconiaceae), a key taxon for understanding the evolution of flower development in the Zingiberales. *American Journal of Botany*, 96, 580–593.

Kitazawa, M. and Fujimoto, K. (2015). A dynamical phyllotaxis model to determine floral organ number. *PLOS Computational Biology*, 11, e1004145.

Kitazawa, M.S. and Fujimoto, K. (2016a). Relationship between the species-representative phenotype and intraspecific variation in Ranunculaceae floral organ and Asteraceae flower numbers. *Annals of Botany*, 117, 925–935.

Kitazawa, M.S. and Fujimoto, K. (2016b). Stochastic occurrence of trimery from pentamery in floral phyllotaxis of Anemone (Ranunculaceae). *Acta Societatis Botanicorum Poloniae*, 85, 3530.

Kitazawa, M.S. and Fujimoto, K. (2018). Spiral phyllotaxis underlies constrained variation in Anemone (Ranunculaceae) tepal arrangement. *Journal of Plant Research*, 131, 459–468.

Kitazawa, M. and Fujimoto, K. (2020). Perianth phyllotaxis is polymorphic in the basal eudicot Anemone and Eranthis species. *Frontiers in Ecology and Evolution*, 8, 70.

Knuth, P. (1899). *Handbuch der Blütenbiologie*. Engelmann, Leipzig.

Koch, L., Wester, P., Lunau K. (2017). To be on the safe site – Ungroomed spots on the bee's body and their importance for pollination. *PLOS ONE*, 12, 0182522.

Kwiatkowska, D. (2006). Primordium formation at the Arabidopsis shoot apex: Quantitative analysis of surface geometry and growth. *Journal of Experimental Botany*, 57, 571–580.

Lee, H.-W. and Levitov, L.S. (2021). Universality in phyllotaxis: A mechanical theory. *arXiv*, 2101.02652.

Lee, A.K., Gilman, I.S., Srivastav, M., Lerner, A.D., Donoghue, M.J., Clement, W.L. (2021). Reconstructing Dipsacales phylogeny using Angiosperms353: Issues and insights. *American Journal of Botany*, 108, 1122–1142.

Levitov, L.S. (1991). Energetic approach to phyllotaxis. *Europhysics Letters*, 14, 533–539.

Lin, Y.F., Chen, Y.Y., Hsiao, Y.Y., Shen, C.Y., Hsu, J.L., Yeh, C.M., Mitsuda, N., Ohme-Takagi, M., Liu, Z.J., Tsai, W.C. (2016). Genome-wide identification and characterization of TCP genes involved in ovule development of Phalaenopsis equestris. *Journal of Experimental Botany*, 67, 5051–5066.

Lippman, Z.B., Cohen, O., Alvarez, J.P., Abu-Abied, M., Pekker, I., Paran, I., Eshed, Y., Zamir, D. (2008). The making of a compound inflorescence in tomato and related nightshades. *PLOS Biology*, 6, e288.

LPWG (Legume Phylogeny Working Group) (2013). Legume phylogeny and classification in the 21s century: Progress, prospects and lessons for other species-rich clades. *Taxon*, 62, 217–248.

LPWG (Legume Phylogeny Working Group) (2017). A new subfamily classification of the Leguminosae based on a taxonomically comprehensive phylogeny. *Taxon*, 66, 44–77.

Lucibelli, F., Valoroso, M.C., Aceto, S. (2020). Radial or bilateral? The molecular basis of floral symmetry. *Genes*, 11, 395.

Lucibelli, F., Valoroso, M.C., Theißen, G., Nolden, S., Mondragon-Palomino, M., Aceto, S. (2021). Extending the toolkit for beauty: Differential co-expression of DROOPING LEAF-like and class B MADS-box genes during Phalaenopsis flower development. *Intenational Journal of Molecular Science*, 22, 7025.

Luo, D., Carpenter, R., Vincent, C., Copsey, L., Coen, E. (1996). Origin of floral asymmetry in Antirrhinum. *Nature*, 383, 794–799.

Luo, D., Carpenter, R., Copsey, L., Vincent, C., Clark, J., Coen, E. (1999). Control of organ asymmetry in flowers of Antirrhinum. *Cell*, 99, 367–376.

Madrigal, Y., Alzate, J.F., Pabon-Mora, N. (2017). Evolution and expression patterns of TCP genes in Asparagales. *Frontiers in Plant Science*, 8, 9.

Martinez-Salazar, S., González, F., Alzate, J., Pabón-Mora, N. (2021). Molecular framework underlying floral bilateral symmetry and nectar spur development in Tropaeolum, an atypical member of the Brassicales. *American Journal of Botany*, 108, 1315–1330.

Marzec, C. and Kappraff, J. (1983). Properties of maximal spacing on a circle related to phyllotaxis and to the golden mean. *Journal of Theoretical Biology*, 103, 201–226.

Meicenheimer, R.D. (1979). Relationships between shoot growth and changing phyllotaxy of Ranunculus. *American Journal of Botany*, 66, 557–569.

Meicenheimer, R.D. (2006). Stem unit growth analysis of *Linum usitatissimum* internode development. *American Journal of Botany*, 93, 55–63.

Naghiloo S. (2020). Patterns of symmetry expression in angiosperms: Developmental and evolutionary lability. *Frontiers in Ecology and Evolution*, 8, 104.

Naghiloo, S. and Claßen-Bockhoff, R. (2017). Developmental changes in time and space promote evolutionary diversification of flowers: A case study in Dipsacoideae. *Frontiers in Plant Science*, 8, 1665.

Naghiloo, S., Dadpour, M.R., Gohari, G., Endress, P.K. (2013). Comparative study of inflorescence development in Oleaceae. *American Journal of Botany*, 100, 647–663.

Nakagawa, A., Kitazawa, M.S., Fujimoto, K. (2020). A design principle for floral organ number and arrangement in flowers with bilateral symmetry. *Development*, 147, dev182907.

Nicolas, M. and Cubas, P. (2016). The role of tcp transcription factors in shaping flower structure, leaf morphology, and plant architecture. In *Plant Transcription Factors*, Gonzalez, D.H. (ed.). Academic Press, Boston.

Nisoli, C., Gabor, N.M., Lammert, P.E., Maynard, J.D., Crespi, V.H. (2009). Static and dynamical phyllotaxis in a magnetic cactus. *Physical Review Letters*, 102, 186103.

O'Meara, B.C., Smith, S.D., Armbruster, W.S., Harder, L.D., Hardy, C.R., Hileman, L.C., Hufford, H., Litt, A., Magallón, S., Smith, S.A. et al. (2016). Non-equilibrium dynamics and floral trait interactions shape extant angiosperm diversity. *Proceedings of the Royal Society B*, 283, 20152304.

Pabón-Mora, N. and Gonzalez, F. (2008). Floral ontogeny of Telipogon spp. (Orchidaceae) and insights on the perianth symmetry in the family. *International Journal of Plant Sciences*, 169, 1159–1173.

Pabón-Mora, N., Madrigal, Y., Alzate, J.F., Ambrose, B.A., Ferrándiz, C., Wanke, S., Neinhuis, C., González, F. (2020). Evolution of Class II TCP genes in perianth bearing Piperales and their contribution to the bilateral calyx in Aristolochia. *New Phytologist*, 228, 752–769.

Pan, Z.J., Nien, Y.C., Shih, Y.A., Chen, T.Y., Lin, W.D., Kuo, W.H., Hsu, H.C., Tu, S.L., Chen, J.C., Wang, C.N. (2022). Transcriptomic analysis suggests auxin regulation in dorsal-ventral petal asymmetry of wild progenitor *Sinningia speciosa*. *International Journal of Molecular Science*, 23, 2073.

Pang, H.B., Pang Sun, Q.W., He, S.Z., Wang, Y.Z. (2010). Expression pattern of CYC-like genes relating to a dorsalized actinomorphic flower in Tengia (Gesneriaceae). *Journal of Systematics and Evolution*, 48, 309–317.

Papadopulos, A.S.T., Powell, M.P., Pupulin, F., Warner, J., Hawkins, J.A., Salamin, N., Chittka, L., Williams, N.H., Whitten, W.M., Loader, D., Valente, L.M., Chase, M.W., Savolainen, V. (2013). Convergent evolution of floral signals underlies the success of Neotropical orchids. *Philosophical Transactions of the Royal Society London B*, 280, 0960.

Payer, J-B. (1857). *Traité d'organogénie végétale comparée de la fleur*. Victor Masson, Paris.

Plantefol, L. (1949). *L'ontogénie de la fleur*. Masson, Paris.

Prenner, G. (2004). New aspects in floral development of Papilionoideae: Initiated but suppressed bracteoles and variable initiation of sepals. *Annals of Botany*, 93, 537–545.

Preston, J.C. and Hileman, L.C. (2012). Parallel evolution of TCP and B-class genes in Commelinaceae flower bilateral symmetry. *EvoDevo*, 3, 6.

Preston, J.C., Kost, M.A., Hileman, L.C. (2009). Conservation and diversification of the symmetry developmental program among close relatives of snapdragon with divergent floral morphologies. *New Phytologist*, 182, 751–762.

Preston, J.C., Martinez, C.C., Hileman, L.C. (2011). Gradual disintegration of the floral symmetry gene network is implicated in the evolution of a wind-pollination syndrome. *The Proceedings of the National Academy of Sciences*, 108, 2343–2348.

Ramírez-Domenech, J.I. and Tucker, S.C. (1989). Phylogenetic implications of inflorescence and floral ontogeny of Mimosa strigillosa. *American Journal of Botany*, 76, 1583–1593.

Rayleigh, J.W. (1916). On convection currents in a horizontal layer of fluid, when the higher temperature is on the under side. *The London, Edinburgh, and Dublin Philosophical Magazine and Journal of Science*, 32, 529–546.

Reinhardt, D. and Gola, E.M. (2022). Law and order in plants – The origin and functional relevance of phyllotaxis. *Trends in Plant Science*, 27, 1017–1032.

Remizowa, M.V., Sokoloff, D.D., Rudall, P.J. (2006). Patterns of floral structure and orientation in Japonolirion, Narthecium, and Tofieldia. *Aliso*, 22, 159–171.

Remizowa, M.V., Sokoloff, D.D., Kondo, K. (2008). Floral evolution in the monocot family Nartheciaceae (Dioscoreales): Evidence from anatomy and development in Metanarthecium luteo-viride Maxim. *Botanical Journal of the Linnean Society*, 158, 1–18.

Remizowa, M.V., Sokoloff, D.D., Kondo, K. (2010). Early flower and inflorescence development in Dioscorea tokoro (Dioscoreales): Shoot chirality, handedness of cincinni and common tepal-stamen primordia. *Wulfenia*, 17, 77–97.

Remizowa, M.V., Rudall, P.J., Choob, V.V., Sokoloff, D.D. (2013). Racemose inflorescences of monocots: Structural and morphogenetic interaction at the flower/inflorescence level. *Annals of Botany*, 112, 1553–1566.

Ren, Y., Hong-Li, C., Endress P. (2010). Floral development in Anemoneae (Ranunculaceae). *Botanical Journal of the Linnean Society*, 162, 77–100.

Ren, Y., Gu, T.-Q., Chang H.-L. (2011). Floral development of Dichocarpum, Thalictrum, and Aquilegia (Thalictroideae, Ranunculaceae). *Plant Systematics and Evolution*, 292, 203–213.

Reyes, E., Sauquet, H., Nadot, S. (2016). Perianth symmetry changed at least 199 times in angiosperm evolution. *Taxon*, 65, 945–964.

Richards, F.J. (1951). Phyllotaxis: Its quantitative expression and relation to growth in the apex. *Philosophical Transactions of the Royal Society London B*, 235, 509–564.

Ridley, J.N. (1982). Packing efficiency in sunflower heads. *Mathematical Biosciences*, 58, 129–139.

Roels, P. and Smets, E.F. (1996). A floral ontogenetic study in the Dipsacales. *International Journal of Plant Sciences*, 157, 203–218.

Ronse de Craene, L.P. (2016). Development and evolution of extreme synorganization in angiosperm flowers and diversity: A comparison of Apocynaceae and Orchidaceae. *Annals of Botany*, 117, 749–767.

Ronse de Craene, L.P. (2017). Floral development of Berberidopsis beckleri – Can an additional species of the Berberidopsidaceae add evidence to floral evolution in the core eudicots? *Annals of Botany*, 119, 599–610.

Ronse de Craene, L.P. (2022). *Floral Diagrams. An Aid to Understanding Flower Morphology and Evolution*. Cambridge University Press, Cambridge.

Ronse de Craene, L.P. and Smets, E. (1990). The systematic relationship between Begoniaceae and Papaveraceae: A comparative study of their floral development. *Bulletin du Jardin Botanique National de Belgique*, 60, 229–273.

Ronse de Craene, L.P. and Smets, E.F. (2001). Floral developmental evidence for the systematic relationships of Tropaeolum (Tropaeolaceae). *Annals of Botany*, 88, 879–892.

Rothen, F. and Koch, A.-J. (1989). Phyllotaxis, or the properties of spiral lattices. Part I: Shape invariance under compression. *Journal de physique*, 50, 633–657.

Rudall, P.J. and Bateman, R.M. (2004). Evolution of zygomorphy in monocot flowers, iterative patterns and developmental constraints. *New Phytologist*, 162, 25–44.

Rutishauser, R. (1998). Plastochron ratio and leaf arc as parameters of a quantitative phyllotaxis analysis in vascular plants. In *Symmetry in Plants*, Jean R.J. and Barabé D. (eds). World Scientific, Singapore.

Rutishauser, R. (2016). *Acacia* (wattle) and *Cananga* (ylang-ylang): From spiral to whorled and irregular (chaotic) phyllotactic patterns – A pictorial report. *Acta Societatis Botanicorum Poloniae*, 85, 3531.

Rutishauser, R. (2019). Ever since Darwin: Why plants are important for evo-devo research. In *Perspectives on Evolutionary and Developmental Biology: Essays for Alessandro Minelli*, Fusco, G. (ed.). Padova University Press, Padova.

Sargent, R.D. (2004). Floral symmetry affects speciation rates in angiosperms. *Proceedings of the Royal Society B*, 271, 603–608.

Schöffel, K. (1932). Untersuchungen über den Blütenbau der Ranunculaceen. *Planta*, 17, 315–371.

Schwendener, S. (1878). *Mechanische Theorie der Blattstellungen*. Engelmann, Leipzig.

Smith, R., Prusinkiewicz, P., Kuhlemeier, C. (2006). Inhibition fields for phyllotactic pattern formation: A simulation study. *Canadian Journal of Botany*, 84, 1635–1649.

Smyth, D.R., Bowman, J.L., Meyerowitz, E.M. (1990). Early flower development in Arabídopsis. *The Plant Cell*, 2, 755–767.

Snow, M. and Snow, R. (1952). Minimum area and leaf determination. *Proceedings of the Royal Society of London B*, 139, 545–566.

Sokoloff, D.D., Remizowa, M.V., Bateman, R.M., Rudall, P.J. (2018). Was the ancestral angiosperm flower whorled throughout? *American Journal of Botany*, 105, 5–15.

Soltis, D.E. and Hufford, L. (2002). Ovary position diversity in Saxifragaceae: Clarifying the homology of epigyny. *International Journal of Plant Sciences*, 163, 277–293.

Soltis, D.E., Soltis, P., Endress, P., Chase, M., Manchester, S., Judd, W., Majure, L., Mavrodiev, E. (2018). *Phylogeny and Evolution of the Angiosperms*. The University of Chicago Press, Chicago.

Souer, E., van der Krol, A.R., Kloos, D., Spelt, C., Bliek, M., Mol, J., Koes, R. (1998). Genetic control of branching pattern and floral identity during Petunia inflorescence development. *Development*, 125, 733–742.

Specht, C.D. and Howarth, D.G. (2015). Adaptation in flower form: A comparative evodevo approach. *New Phytologist*, 206, 74–90.

Spencer, V. and Kim, M. (2018). Re"CYC"ling molecular regulators in the evolution and development of flower symmetry. *Seminars in Cell and Developmental Biology*, 79, 16–26.

Staedler, Y. and Endress, P.K. (2009). Diversity and lability of floral phyllotaxis in the pluricarpellate families of core Laurales. *International Journal of Plant Sciences*, 170, 522–550.

Staedler, Y., Weston, P., Endress, P.K. (2007). Floral phyllotaxis and floral architecture in Calycanthaceae (Laurales). *International Journal of Plant Sciences*, 168, 285–306.

Strauss, S., Lempe, J., Prusinkiewicz, P., Tsiantis, M., Smith, R.S. (2020). Phyllotaxis: Is the golden angle optimal for light capture? *New Phytologist*, 225, 499–510.

Thomé, O.W. (1885). *Flora von Deutschland, Österreich und der Schweiz*. Gera, Germany.

Tucker, S.C. (1989). Overlapping organ initation and common primordia in flowers of *Pisum sativum* (Leguminosae: Faboideae). *American Journal of Botany*, 76, 714–729.

Tucker, S.C. (1996). Trends in evolution of floral ontogeny in Cassia sensu stricto, Senna, and Chamaecrista (Leguminosae: Caesalpinioideae: Cassieae: Cassiinae); A study in convergence. *American Journal of Botany*, 83, 687–711.

Tucker, S.C. (1997). Floral evolution, development, and convergence: The hierarchical-significance hypothesis. *International Journal of Plant Sciences*, 158, S143–S161.

Tucker, S.C. (2002a). Floral ontogeny of Cercis (Leguminosae: Caesalpinioideae: Cercideae): Does it show convergence with papilionoids? *International Journal of Plant Sciences*, 163, 75–87.

Tucker, S.C. (2002b). Comparative floral ontogeny in Detarieae (Leguminosae: Caesalpinioideae). 1. Radially symmetrical taxa lacking organ suppression. *American Journal of Botany*, 89, 875–887.

Tucker, S.C. (2002c). Floral ontogeny in Sophoreae (Leguminosae: Faboideae). III. Radial symmetry and random petal aestivation in Cadia purpurea. *American Journal of Botany*, 89, 748–757.

Tucker, S.C. (2003). Floral development in legumes. *Plant Physiology*, 131, 911–926.

Tucker S.C. and Hodges S. (2005). Floral ontogeny of Aquilegia, Semiaquilegia and Enemion (Ranunculaceae). *International Journal of Plant Sciences*, 166, 557–674.

Vamosi, J.C., Magallon, S., Mayrose, I., Otto, S.P., Sauquet, H. (2018). Macroevolutionary patterns of flowering plant speciation and extinction. *Annual Review of Plant Biology*, 69, 685–706.

Vincent, C.A. and Coen, E.S. (2004). A temporal and morphological framework for flower development in *Antirrhinum majus*. *Canadian Journal of Botany*, 82, 681–690.

Vöchting, H. (1885). Über die Ursachen der Zygomorphie der Blüthen. *Berichte der Deutschen Botanischen Gesellschaft*, 3, 341–345.

Vogel, S. (1990). History of the Malpighiaceae in the light of pollination ecology. *Memoirs of the New York Botanical Garden*, 55, 130–142.

Walch, J.-P. (2022). Modeling phyllotaxis: From the inhibition potential to the real plant. *Journal of Theoretical Biology*, 553, 111261.

Walch, J.-P. and Blaise, S. (2022a). Modeling basal angiosperm floral phyllotaxis. *Botany Letters*, 169, 305–318.

Walch, J.-P. and Blaise S. (2022b). Mathematical modeling the influence of bracteoles on sepal arrangement. *Flora*, 295, 152138.

Walch, J.-P. and Blaise, S. (2023). *Phyllotaxis Models: A Tool for Evolutionary Biologists*. ISTE Ltd, London, and John Wiley & Sons, New York.

Walch, J.-P. and Blaise S. (2024). Perianth phyllotaxis and symmetry in Lamiales. *Botany Letters*, 1–5.

Wang, Z., Luo, Y., Li, X., Wang, L., Xu, S., Yang, J., Weng, L., Sato, S., Tabata, S., Ambrose, M. et al. (2008). Genetic control of floral zygomorphy in pea (*Pisum sativum*). *The Proceedings of the National Academy of Sciences*, 105, 10414–10419.

Weisse, A. (1894). Neue Beitraäge zur mechanischen Blattstellungslehre. *Jahrbücher für Wissenschaftliche Botanik*, 26, 236–294.

Welty, N., Radovich, C., Meulia, T., van der Knaap, E. (2007). Inflorescence development in two tomato species. *Canadian Journal of Botany*, 85, 111–118.

Wessinger, C.A. and Hileman, L.C. (2020). Parallelism in flower evolution and development. *Annual Review of Ecology, Evolution and Systematics*, 51, 387–408.

Whittall, J.B. and Hodges, S.A. (2007). Pollinator shifts drive increasingly long nectar spurs in columbine flowers. *Nature Letters*, 447, 406–409.

Wiegel, M. and Lunau, K. (2023). Flower symmetry in Saxifraga revisited. *Plant Systematics and Evolution*, 309, 6.

Williams, R. (1975). *The Shoot Apex and Leaf Growth*. Cambridge University Press, Cambridge.

Winfield, P. (1987). A new gene (lst) which influences standard shape in Pisum sativum. *Pisum Newsletter*, 19, 84–85.

Wiss, D. and Zagórska-Marek, B. (2012). Geometric parameters of the apical meristem and the quality of phyllotactic patterns in Magnolia flowers. *Acta Societatis Botanicorum Poloniae*, 81, 203–216.

Yin, X. and Meicenheimer, D. (2017). The ontogeny, phyllotactic diversity, and discontinuous transitions of *Diphasiastrum digitatum* (Lycopodiaceae). *American Journal of Botany*, 104, 8–23.

Yu, Q., Tian, X., Lin, C., Specht, C.D., Liao, J. (2020). Expression and function studies of CYC/TB1-like genes in the asymmetric flower Canna (Cannaceae, Zingiberales). *Frontiers in Plant Science*, 11, 580576.

Yuan, Z., Gao, S., Xue, D.-W., Luo, D., Li, L.-T., Ding, S.-Y., Yao, X., Wilson, Z.A., Qian, Q., Zhang, D.-B. (2009). RETARDED PALEA1 controls palea development and floral zygomorphy in rice. *Plant Physiology*, 149, 235–244.

Zagórska-Marek, B. (1987). Phyllotaxis triangular unit: Phyllotactic transitions as the consequence of apical wedge disclinations in a crystal-like pattern of the units. *Acta Societatis Botanicorum Poloniae*, 56, 229–255.

Zagórska-Marek, B. (1994). Phyllotaxic diversity in Magnolia flowers. *Acta Societatis Botanicorum Poloniae*, 63, 117–137.

Zhang, T. and Elomaa, P. (2021). Don't be fooled: False flowers in Asteraceae. *Current Opinion in Plant Biology*, 59, 101972.

Zhang, X.H. and Ren, Y. (2011). Comparative floral development in Lardizabalaceae (Ranunculales). *Botanical Journal of the Linnean Society*, 166, 171–184.

Zhang, W., Kramer, E.M., Davis, C.C. (2010). Floral symmetry genes and the origin and maintenance of zygomorphy in a plant-pollinator mutualism. *The Proceedings of the National Academy of Sciences*, 107, 14.

Zhang, W., Steinmann, V.W., Nikolov, L., Kramer, E.M., Davis, C.C. (2013). Divergent genetic mechanisms underlie reversals to radial floral symmetry from diverse zygomorphic flowered ancestors. *Frontiers in Plant Science*, 4, 302.

Zhang, J., Stevens, P., Zhang, W. (2017). Evolution of floral zygomorphy in androecium and corolla in Solanaceae. *Journal of Systematics and Evolution*, 55, 581–590.

Zhang, J., Stevens, P., Zhang, W. (2022a). Evolution and development of inflorescences and floral symmetry in Solanaceae. *American Journal of Botany*, 109, 746–767.

Zhang, W.G., Liu, Y.G., Nie, T.J., Guo, C.C., Qiu, L.H., Yang, G.Y., Jabbour, F., Wang, W. (2022b). Floral ontogeny of *Delphinium anthriscifolium* (Ranunculaceae) and development of intrafloral and symmetrical resupinated organs. *Botanical Journal of the Linnean Society*, 198, 86–98.

Zhao, L., Liu, P., Che, X.F., Wang, W., Ren, Y. (2011). Floral organogenesis of *Helleborus thibetanus* and *Nigella damascena* (Ranunculaceae) and its systematic significance. *Botanical Journal of the Linnean Society*, 166, 431–443.

Zhao, L., Bachelier, J.B., Chang, H.L., Tian, X.H., Ren, Y. (2012). Inflorescence and floral development in Ranunculus and three allied genera in Ranunculeae (Ranunculoideae, Ranunculaceae). *Plant Systematics and Evolution*, 298, 1057–1071.

Zhao, Y., Pfannebecker, K., Dommes, A.B., Hidalgo, O., Becker, A., Elomaa, P. (2018). Evolutionary diversification of CYC/TB1-like TCP homologs and their recruitment for the control of branching and floral morphology in Papaveraceae (basal eudicots). *New Phytologist*, 220, 317–331.

Zhao, Y., Zhang, R., Jiang, K.W., Qi, J., Hu, Y., Guo, J., Zhu, R., Zhang, T., Egan, A.N., Yi, T.S., Huang, C.H., Ma, H. (2021). Nuclear phylotranscriptomics and phylogenomics support numerous polyploidization events and hypotheses for the evolution of rhizobial nitrogen-fixin symbiosis in Fabaceae. *Molecular Plant*, 14, 748–773.

Zhao, H., Liao, H., Li, S., Zhang, R., Dai, J., Ma, P., Wang, T., Wang, M., Yuan Y., Fu, X. et al. (2023). Delphinieae flowers originated from the rewiring of interactions between duplicated and diversified floral organ identity and symmetry genes. *The Plant Cell*, 35, 994–1012.

Zhao-Jun, P., Nien, Y.-C., Shih, Y.-A., Chen, T.-Y., Lin, W.-D., Kuo, W.-H., Hsu, H.-C., Tu, S.-L., Chen, J.-C., Wang, C.-N. (2022). Transcriptomic analysis suggests auxin regulation in dorsal-ventral petal asymmetry of wild progenitor *Sinningia speciosa*. *International Journal of Molecular Science*, 23, 2073.

Zhong, J. and Kellogg, E.A. (2015). Stepwise evolution of corolla symmetry in CYCLOIDEA2-like and RADIALIS-like gene expression patterns in Lamiales. *American Journal of Botany*, 102, 1260–1267.

Zhong, J., Preston, J.C., Hileman, L.C., Kellogg, E.A. (2017). Repeated and diverse losses of corolla bilateral symmetry in the Lamiaceae. *Annals of Botany*, 119, 1211–1223.

Zhou, X.R., Wang, Y.Z., Smith, J.F., Chen, R. (2008). Altered expression patterns of TCP and MYB genes relating to the floral developmental transition from initial zygomorphy to actinomorphy in Bournea (Gesneriaceae). *New Physiologist*, 178, 532–543.

Zoulias, N., Duttke, S.H.C., Garcês, H., Spencer, V., Kim, M. (2019). The role of auxin in the pattern formation of the Asteraceae flower head (capitulum). *Plant Physiology*, 179, 391–401.

Index

A, B

C

D, E

F, H, I

L, O, P

R, S, T

U, W, Z

Other titles from

ISTE

in

Biology and Biomedical Engineering

2024

MASQUIDA Benoît, LECLERC Fabrice
Looking at Ribozymes: Biology of Catalytic RNA
(Nucleic Acids Set – Volume 2)

2023

BALTHAZART Jacques
Fifty Years of Evolution in Biological Research: Progress and Decline

GILBERT Marc, PIRKMAJER Sergej
Concepts in Biology: A Historical Perspective

VERNA Emeline
Asymptomatic Osseous Variations of the Postcranial Human Skeleton
(Comparative Anatomy and Posture of Animal and Human Set – Volume 5)

WALCH Jean-Paul, BLAISE Solange
Phyllotaxis Models: A Tool for Evolutionary Biologists

2022

DAMBRICOURT MALASSÉ Anne
Embryogeny and Phylogeny of the Human Posture 2: A New Glance at the Future of our Species
(Comparative Anatomy and Posture of Animal and Human Set – Volume 4)

FOSSÉ Philippe
Structures and Functions of Retroviral RNAs: The Multiple Facets of the Retroviral Genome
(Nucleic Acids Set – Volume 1)

2021

DAMBRICOURT MALASSÉ Anne
Embryogeny and Phylogeny of the Human Posture 1: A New Glance at the Future of our Species
(Comparative Anatomy and Posture of Animal and Human Set – Volume 3)

GRANDCOLAS Philippe, MAUREL Marie-Christine
Systematics and the Exploration of Life

HADJOUIS Djillali
The Skull of Quadruped and Bipedal Vertebrates: Variations, Abnormalities and Joint Pathologies
(Comparative Anatomy and Posture of Animal and Human Set – Volume 2)

HULLÉ Maurice, VERNON Philippe
The Terrestrial Macroinvertebrates of the Sub-Antarctic Îles Kerguelen and Île de la Possession

2020

CAZEAU Cyrille
Foot Surgery Viewed Through the Prism of Comparative Anatomy: From Normal to Useful
(Comparative Anatomy and Posture of Animal and Human Set – Volume 1)

DUJON Bernard, PELLETIER Georges
Trajectories of Genetics

2019

BRAND Gérard
Discovering Odors

BUIS Roger
Biology and Mathematics: History and Challenges

2016

CLERC Maureen, BOUGRAIN Laurent, LOTTE Fabien
Brain–Computer Interfaces 1: Foundations and Methods
Brain–Computer Interfaces 2: Technology and Applications

FURGER Christophe
Live Cell Assays: From Research to Health and Regulatory Applications

2015

CLARYSSE Patrick, FRIBOULET Denis
Multi-modality Cardiac Imaging: Processing and Analysis

2014

CHÈZE Laurence
Kinematic Analysis of Human Movement

DAO Tien Tuan, HO BA THO Marie-Christine
Biomechanics of the Musculoskeletal System: Modeling of Data Uncertainty and Knowledge

FANET Hervé
Medical Imaging Based on Magnetic Fields and Ultrasounds

FARIÑAS DEL CERRO Luis, INOUE Katsumi
Logical Modeling of Biological Systems

MIGONNEY Véronique
Biomaterials

TEBBANI Sihem, LOPES Filipa, FILALI Rayen, DUMUR Didier,
PAREAU Dominique
CO_2 Biofixation by Microalgae: Modeling, Estimation and Control

Printed and bound by CPI Group (UK) Ltd, Croydon, CR0 4YY
16/09/2026
14963089-0001